Crisis Urbanism and Postcolonial African Cities in Postmillennial Cinema

This book provides a framework to rethink postcoloniality and urbanism from African perspectives. Bringing together multidisciplinary perspectives on African crises through postmillennial films, the book addresses the need to situate global south cultural studies within the region.

The book employs film criticism and semiotics as devices to decode contemporary cultures of African cities, with a specific focus on crisis. Drawing on a variety of contemporary theories on cities of the global south, especially Africa, the book sifts through nuances of crisis urbanism within postmillennial African films. In doing so the book offers unique perspectives that move beyond the confines of sociological or anthropological studies of cities. It argues that crisis has become a mainstay reality of African cities and thus occupies a central place in the way these cities may be theorized or imagined. The book considers crises of six African cities: nonentity in post-apartheid Johannesburg, laissez faire economies of Kinshasa, urban commons in Nairobi, hustlers in postwar Monrovia, latent revolt in Cairo, and cantonments in postwar Luanda, which offer useful insights on African cities today.

The book will be of interest to students and scholars of urban studies, urban geography, urban sociology, cultural studies, and media studies.

Addamms Mututa is a postdoctoral research fellow at the African Centre for Cities, University of Cape Town. He practices interdisciplinary research on cinema and global south cities and holds a joint PhD from the University of Tübingen, Germany, and the University of the Witwatersrand, Johannesburg.

Routledge Research on Decoloniality and New Postcolonialisms

Series Editor: Mark Jackson, Senior Lecturer in Postcolonial Geographies, School of Geographical Sciences, University of Bristol, UK.

Routledge Research on Decoloniality and New Postcolonialisms is a forum for original, critical research into the histories, legacies, and life-worlds of modern colonialism, postcolonialism, and contemporary coloniality. It analyses efforts to decolonise dominant and damaging forms of thinking and practice, and identifies, from around the world, diverse perspectives that encourage living and flourishing differently. Once the purview of a postcolonial studies informed by the cultural turn's important focus on identity, language, text, and representation, today's resurgent critiques of coloniality are also increasingly informed, across the humanities and social sciences, by a host of new influences and continuing insights for different futures: indigeneity, critical race theory, relational ecologies, critical semiotics, post-humanisms, ontology, affect, feminist standpoints, creative methodologies, post-development, critical pedagogies, intercultural activisms, place-based knowledges, and much else. The series welcomes a range of contributions from socially engaged intellectuals, theoretical scholars, empirical analysts, and critical practitioners whose work attends, and commits, to newly rigorous analyses of alternative proposals for understanding life and living well on our increasingly damaged earth.

This series is aimed at upper-level undergraduates, research students, and academics, appealing to scholars from a range of academic fields including human geography, sociology, politics and broader interdisciplinary fields of social sciences, arts, and humanities.

A Decolonial Black Feminist Theory of Reading and Shade
Feeling the University
Andrea N. Baldwin

Transdisciplinary Thinking from the Global South
Whose problems, whose solutions?
Edited by Juan Carlos Finck Carrales and Julia Suárez-Krabbe

For more information about this series, please visit: https://www.routledge.com/Routledge-Research-on-Decoloniality-and-New-Postcolonialisms/book-series/RRNP

Crisis Urbanism and Postcolonial African Cities in Postmillennial Cinema

Addamms Mututa

Routledge
Taylor & Francis Group

LONDON AND NEW YORK

First published 2022
by Routledge
2 Park Square, Milton Park, Abingdon, Oxon OX14 4RN

and by Routledge
605 Third Avenue, New York, NY 10158

Routledge is an imprint of the Taylor & Francis Group, an informa business

British Library Cataloguing-in-Publication Data
A catalogue record for this book is available from the British Library

Library of Congress Cataloging-in-Publication Data
A catalog record has been requested for this book

ISBN: 978-0-367-64083-5 (hbk)
ISBN: 978-0-367-64085-9 (pbk)
ISBN: 978-1-003-12209-8 (ebk)

DOI: 10.4324/9781003122098

Typeset in Times New Roman
by SPi Technologies India Pvt Ltd (Straive)

To my late father and mother. All is well.

Contents

Acknowledgements   *viii*

1   Crisis urbanism in context   1

2   Crisis of nonentity: "unknowing" Johannesburg's
    post-apartheid townships   21

3   *Laissez faire* urbanism: economies of dystopia
    in postcolonial Kinshasa   39

4   Urbanism of the commons: inhabiting trash and
    a crisis of communing in Nairobi   58

5   Rarray urbanism: the superficies of Monrovia's
    hustlers in postwar urban crisis   76

6   Revolt urbanism: Cairo's crisis citizenship under
    construction   96

7   Outlier urbanism: inside Luanda's postwar
    cantonments   113

8   Crisis urbanism and the future of African cities   129

Index   *138*

Acknowledgements

This book has benefitted from the generous support of various individuals and organizations. I acknowledge the immense support received from Professor Edgar Pieterse – chair of African Centre for Cities (ACC) at the University of Cape Town. I am deeply indebted to you for your intellectual input and generous material support during the proofreading of the book manuscript. In this regard, I am also very grateful to colleagues at ACC for enabling the process and to Lee Middleton for her excellent job of proofreading the manuscript. I also thank ACC for offering me the AW Mellon Postdoctoral Research Fellowship in Southern Urbanism which enabled me to write this manuscript. Without this opportunity and support, the task would have been nearly impossible. I appreciate Professor Dr. Russell West-Pavlov, co-convenor of the Interdisciplinary Centre for Global South Studies at the University of Tübingen, and the late Professor Bhekizizwe Peterson, formerly of African Literature department at the University of the Witwatersrand, for excellent intellectual mentorship as my doctoral supervisors. I thank Professor Ndeti Ndati of the University of Nairobi for his kind support that has continued to date. Finally, I appreciate my family for the kind support in this process.

1 Crisis urbanism in context

Introduction

Global south cities are generally known for their chaos, high population density, violence, informality, crime, and other comparable characteristics (De Boeck and Plissart 2004; Foster 2009; Gastrow 2017; Gastrow 2018; Murray 2008, 2011). Postcolonial African cities, exemplary of global south cities, are particularly perceived as "'dangerous', chaotic, dysfunctional and non-productive" (Locatelli and Nugent 2009, 2). Yet, despite any vulnerability or disorientation that a stranger may feel when first encountering these cities, to many residents, such crises are barely perceptible as they have become commonplace experiences. It is evident that surviving the crisis produced by such ostensibly crushing chaos and danger has become a popular – if not almost always the only possible – form of urbanism. This realization has the potential to alter the way we think of African cities beyond the corpus of existing scholarship, which problematizes these urban conditions of crisis as markers of degenerative urbanism. To the contrary, we would regard crisis and the responses that it precipitates as emblematic of the practicality of everyday life in global south cities. This book, *Crisis Urbanism and Postcolonial African Cities in Postmillennial Cinema*, attempts to advance this alternative thinking. Referring to the responses that make it possible to survive such everyday precarity in postcolonial African cities as "crisis urbanism," this book deviates from the long tradition of comparative reading of global south cities, instead offering crisis urbanism as a counter-theory and conceptual method to theorize global south cities.

The perspective on crisis adopted in this book is not alarmist or dystopian. Noting that Africa's urbanism, unlike that in the developed world, offers unique difficulties to urban dwellers, the book uses crisis as a theoretical lens to respond to the questions: what perspectives on urban life do the choices made by Africa's urban populations, particularly in moments of intense difficulties, enable? What forms of urbanism do the reactions to various crises in African cities make possible? Through which local lens are we to read such urbanism in the context of the creative imaginaries it motivates in creative arts like films? The point is thus not that African cities are dystopic but that intense difficulties motivate unusual urban practices that themselves elicit conversations about the meaning of urban citizenship in postcolonial cities

DOI: 10.4324/9781003122098-1

through the general application of crises as an umbrella theoretical perspective. It is noteworthy that the book does not frame crisis as an escapable scenario. It frames it as a dynamic mixture of events, periods, moments, people, spaces, politics, and all sorts of other urban elements which, by their presence, action, or being acted upon, dictate the undertakings and opportunities that are possible in any given day and hour. Put differently, crisis is seen as a condition that necessitates new sets of urban skills and practices, thus playing a central role in designating how cities evolve. For the case of Africa, crisis urbanism theory occupies the disjuncture between global north-south urban theories.

Global north-global south urban theory: a disjuncture

In his academic seminar on Epistemological Practices of Southern Urbanism at African Centre for Cities (ACC) at the University of Cape Town, Edgar Pieterse (2014) grapples with the question: "how best can meaningful urban knowledge be produced?". Particularly, he seeks a

> Research practice that can resist the demands for definite conclusions and solutions, resist the temptation for generalized abstraction about 'the poors' or the subalterns, and rather settle for a careful elucidation of the many folds and intimacies regarding processes of subjectification.
>
> (Ibid., 4)

In an era of disciplinary convergence in global urban studies, Pieterse's discussions pre-empt a critical acknowledgement: that a more connected, supple, and intuitive process of producing urban knowledge is both overdue and welcome. Some of the global south urban theorists whose works have sought new entry points into theorizing the global south include Kees Koonings and Dirk Kruijt (2009), Susan Parnell and Sophie Oldfield (2014), Leigh Anne Duck (2015), and Richard De Satge and Vanessa Watson (2018), among others.

Yet, from a pragmatic standpoint, such efforts can be positioned within an ever-increasing gap that exists due to the limits of applying global north urban theories in the interpretation of global south cities. As the pursuit of new theories with which to discuss the world's metropolitan realities is rapidly shifting towards a south-centred interpretation of cities (Comaroff and Comaroff 2014; Parnell and Oldfield 2014), there remains an urgent "need to move towards a more dispassionate approach to the real city, the real economy and the real social practices and identities of the majority of urbanites who are building our cities" (Pieterse 2011, 14). Further, Pieterse advocates for a "much more layered theoretical framework that can foreground the specificity of spatial practices in our diverse cities and towns" (Ibid.). From here, an argument is possible that we cannot sufficiently understand southern urbanism if we approach it "according to Western criteria of functionality," or treat it as a "derivative or a *doppelganger*, a callow copy or a counterfeit, of the Euro-American 'original'" (Locatelli and Nugent 2009, 2). To the contrary, "it demands to be apprehended and addressed in its own right" (Ibid., 7). Upon scrutiny, what Pieterse seems to petition for are interpretations of global south cities intended not so much to oust western theories, but rather to account for the emergent and widespread peculiarities of – as in the case of this book – the postcolonial African city. Efforts dedicated to initiate new dialogues, perspectives, and ways of seeing African cities – the global south cities par excellence – that go beyond itemizing the insufficiencies of the so-called global north urban theories, are overdue. Whatever discipline(s) or approach(es) best serves this purpose, such a quest should be keenly interested in the peculiarities of Africa's urban conditions and the practical responses enabled by such conditions, both of which are critical aspects of the continent's emergent postcolonial urbanism. To this end, I propose some useful points for consideration.

First, that African cities remain widely misunderstood either through the application of theories from global north cities, or through research that exclusively focuses on globally pertinent areas of interest, such as infrastructure and development, while overlooking the deeper nuances of everyday urban crises in diverse and often unobvious forms. In a region where the city "reveals itself first and foremost through its discontinuities, its provisionality and fugitiveness, its superfluousness" (Mbembe 2008, 64), and where "the 'city' and the symbolic nature of representational aspects of the city's being are more difficult to acknowledge" (Çınar and Bender 2007, xv), this is not a trivial matter. One needs to pay attention to the way the irrational, the mundane, the spontaneous, and the unexpected occurrences can easily become mainstream aspects of Africa's postcolonial urbanism. To understand the contemporary African city is to understand how different phenomena are responding to mounting global pressures (Harding and Blokland 2014, 4), and how a city's blend of humans, spatial deals, and social deals calibrate its cultures (Southall 1998, 4). Contingent upon many happenstances, this reality is neither obvious nor straightforward.

Second, that research on the future implications of these aspects of postcolonial cities remains largely scarce. Nas, de Groot, and Schut (2011, 7) note that social scholars have "ignored the symbolic dimension and its interpretation," while anthropologists "are hardly concerned with the city and especially not the city as a whole." Such bold assertions are provocative as they suggest a significant void in the study of Africa's urbanism where the observed practices are deeply ingrained in underlying symbols. Thus, whereas approaches to the functionality of various urban systems such as settlements, economies, and culture may increase knowledge on those areas (Harding and Blokland 2014, 2), they harbour a glaring inadequacy, namely, subjective and normative analysis of the city. Such studies hardly account for the diverse crises in many urban practices, that is, the "social velocity, the power of the unforeseen and of the *unfolding*" (Mbembe and Nuttall 2004, 349) as a useful perspective on Africa's postcolonial urbanism. The "unforeseen and of the *unfolding*" here is a call to scale up Africa's urban theory to permit other possibilities of seeing the "routine everyday life" as a model of future urbanism that responds to a "permanent 'state of emergency'" (Pieterse 2014, 2). Such everyday emergencies manifest in job scarcity, economic difficulties, social stratifications, lack of infrastructure, political problems, and so on. For this reason, time is ripe to discuss African cities beyond the descriptive analysis of torture, confiscation, encumbrance, displacement, and decay or overemphasis on the struggle, resistance, hope, persistence, and re-imagination of segmental and subverted identities. The new frontier is tasked to account for the ecotones between urban disciplines: in the case of this book, between the findings of anthropological, sociological, political, and urban theories on one hand, and prevalent imaginaries of the kind of cities that have been so studied.

Now comes the question which Pieterse poses in his seminar: what would be a befitting approach to theorize postcolonial African urbanism today? One possible inroad is to be found in the work of Alev Çınar and Thomas Bender (2007, xi) who summarize the process of urban theorization as "generation of a formal definition of the city and a characterization of the urban experience." Further, they see the act of imagination as powerful and sufficient to explicate the ways in which the city is "conceived so unquestionably and obviously as a single space" (Ibid., xii). The approach being suggested here is certainly peculiar, particularly because it contrasts the works of renowned global urban theorists, such as Manuel Castells (1977) and Henri Léfebvre (1991, 1996), who theorized the city as a system whose components are easily accessible to sociological and anthropological inquiry. The other possibility, and this seems to be popular among African city theorists, such as Edgar Pieterse, AbdouMaliq Simone, Susan Parnell, Mamadou Diouf, Sophie Oldfield, Filip De Boeck, and Achille Mbembe, is the imaginative turn. The question of how city users imagine themselves in relation to the cities they occupy is frequently becoming central to urban theorization.

However, a complementary theory based on how other disciplines imagine cities is overdue. This is already manifested in literary studies as well as cinema studies of global south cities. The theory being sought is not a complex intellectualization of modernization and its processes but a sensible theory derived "from a lived praxis, one that may occur anywhere and everywhere" (Comaroff and Comaroff 2014, 20). Such a theory would explain urban life and its contingencies, remain attentive to the inconsistencies of everyday urban life as an inherent paradigm of cities, and also contemplate the role of urban imaginaries in capturing, and indeed articulating, ways of engaging with these inconsistencies as apertures into urban form. AbdouMaliq Simone (2008, 79) already anticipated this shift in his idea of "people as infrastructure," which calls for a focus on the urbanite's "concrete acts and contexts of social collaboration inscribed with multiple identities rather than in overseeing and enforcing modulated transactions among discrete population groups." This can be construed to encompass, *inter alia*, the concept of people-mediated urbanism as the new frontier of theorizing African cities, where transactions and interactions are central to rendering the superficies of postcolonial urbanism. The contemporary southern urbanism theorist must note this shift.

And so, a new motivation to interrogate earlier urban theories – not so much to replace or usurp them but to probe the extent to which their exploration of Africa's peculiar urban condition can be further advanced – becomes a welcome possibility. Parnell and Robinson (2012, 595) anticipate "some fruitful areas of emerging conceptualization and debate that could inform these theoretical shifts, and which might better accommodate the recalibration in urban studies." The earlier theories, seen as expressions of "dialectics of contemporary world history" (Comaroff and Comaroff 2014), are thus open for deconstruction. As the traditional world-binary shifts, and the north evolves southward (Ibid., 18), the African city is now, more than ever, at the forefront of this transition (Mbembe and Nuttall 2008). What seemed, in earlier theories, to be spectacular differences between cities may be, indeed, their biggest similarities, if we operationalize these differences as scalar qualities rather than parametric quantities. Thus, by theorizing urban space and urbanism as qualitative rather than quantitative spaces – just as modernity and modernism are qualitative positions of specific levels of civilization – we arrive at a point where theorization of Africa's urbanism is more focused on the urbanite, urban culture, and urban sensitivities. The task today, then, is to circumvent the normative methods and theories, and seek alternative interpretations of postcolonial urban Africa. Given that for millions of Africa's city residents these aspects unfold informally, spontaneously, and often in moments of urgent distress, crisis urbanism – the theory advanced here – becomes opportune in operationalizing crisis as a force that forges new urban forms.

Building on ongoing efforts to grasp African cities – coming from engineering, sociology, anthropology, political sciences, and urban studies – *Crisis Urbanism and Postcolonial African Cities in Postmillennial Cinema* pursues a cross-disciplinary engagement with Africa's postcolonial urbanism. It particularly explores how urban theories may be brought to a conversation about prevailing urban issues through a critique of crisis within the urban imaginary. This book is not an attempted overhaul of existing urban theories, whether these emanate from the global north or global south. Less so does it claim to suddenly uncover some entry point into theorizing Africa's urbanism that works for all time and circumstances. Rather, it treats existing urban theories as porous, with many interlocking possibilities for further engagement. If it appears that the arguments in this book recant, contradict, or undercut existing restraints typical of north and south institutions (Dirlik 2007, 15), or dramatize the "separation – *between* races, cultures, histories, *within* histories – a separation between *before* and *after* that repeats obsessively the mythical moment or disjunction" (Bhabha 1994, 82) between these two global framings, it does so only coincidentally. Its primary goal is to augment, not disturb, such elegant academic accomplishments. The revisions it anticipates are precisely those laid down by Ali Madani-Pour (1995, 28): that urban spaces should be read to promote understanding of "time-space relationships for the betterment of humankind." This is important for achieving what Parnell and Robinson (2012, 598) consider a decentring of "one of the currently dominant themes in urban theory – critiques of urban neoliberalism." To this end, a trimming, or at least a realignment of prevailing notions of southern urbanism is due, if only to moderate the "ontological claim that Afromodernity exists *sui generis*, not as a derivative of the Euro-original" (Comaroff and Comaroff 2014, 19). As such, the theories sought and generated in this book have been allowed to cross-pollinate, even if only subtly, with existing urban theory – a deliberate choice that allows for a flexible reading of Africa's postcolonial urbanism.

By interweaving urban theories with cinema theories, the book seeks to expand Africa's urban epistemology by way of accessing the "affective functions of popular practices, because it is only through the redeployment of such registers that one can begin to fathom what is going on in the real city, and potentially animate a resonant engagement with the city" (Pieterse 2011, 18). To poke into Africa's urbanism as a form of affective consciousness, then, would mean to interrogate multiple signals centred on human-as-cause or agent of that urbanism. The full implication of this choice will unfold in the subsequent chapters of the book. Thus, although *Crisis Urbanism and Postcolonial African Cities in Postmillennial Cinema* will be read alongside other titles about global south theories, urban literary cultures, and Africa's urbanism – whose arguments it intensifies – its unique value to the reader is that, through the lens of crisis urbanism, it focuses on the urban imaginary as a useful dimension of urban theorization today.

Crisis urbanism

In the context of postcolonial African cities, crisis in many ways shapes the urban form by dictating what forms of urbanisms are possible. This is not a negative thing, as crisis is often exploited to justify unconventional or informal strategies deployed to express views, claim urban space, survive in the city, and so on. In Johannesburg, for instance, building hijacking has not only altered the urban form post-1994 but has become a mainstay of urban life because of the opportunity it offers to the urban poor to stake out a space in the city without the usual prerequisite for economic and political capital (Mututa 2020). That is, despite this crisis – or indeed because of it – thousands of inhabitants have made inroads into the inner city neighbourhood of Hillbrow, a formerly exclusive residential space in the city of Johannesburg. There, they have embraced the area's harsh realities of crime, drugs, prostitution, insecurity, and economic difficulties. Crisis in this case is not so much a marker of hopelessness but rather of moments and scenarios where the African urbanite has to make difficult and critical decisions in order to survive. Such decisions can generate optimism rather than desperation about urban life. The argument pursued throughout this book does not edify a stereotyping of Africa's urban savagery or misery but seeks a theorization responsive to the forms of urbanisms that become possible when urban dwellers in Africa are faced with intense difficulties, and must constantly and sometimes repetitively face such moments of crisis decisively to secure survival. This productive function of crisis, which crosscuts the various discussions of African cities in this book, resonates with other historical conversations on crisis urbanism.

In his book on medieval Rome, Chris Wickham (2015, 386) ends with a discussion of the "changing structural constraints faced by Rome's rulers, and, more widely, Rome's elites – all the city's political players, indeed – when they made political choices." Through a historiography of Rome's judicial institutions, politics, and political economy, Wickham discusses the changing political system and the social issues of the commune, exploring the way urban governance became a form of crisis. This is relatable to Frank Eckardt's (2015, 19) discussion of southern European cities, where he notes: "the changes in attitude towards the role of politics and the state are the starting point of a profound crisis of society." Further, he describes the city as a "place of public drama" (Ibid., 23), that is, the locale of the ongoing crisis. Beyond the crisis of modernity evident in such institutions as urban governance, other aspects of public drama would include technology (Soyata et al. 2019), fiscal economies (Alcaly and Mermelstein 2006), and cultural sustainability (Nadarajah and Yamamoto 2007). Eckardt's (2015, 27) observation – that the "city," the descriptive vocabulary is missing, because it is a bulky, irritating, exhilarating, seductive, and intensive experience of oneself and of others, which you do not want to tame by any language" – will benefit the discussion of crisis urbanism throughout this book. Primarily, this is the case because it calls for a rebellious conceptualization of the city beyond its physical infrastructure, one that "derives more from gestures, looks, acts, mimicry and non-verbal symbols and infinite fictional constructs" (Ibid.).

Hence, studies such as Kuniko Fujita's (2013), which acknowledge the challenges posed by urban crisis to existing urban theories, usefully supplement efforts to expand these theories further, as well as provide a fresh starting point for new approaches to imagining modern cities. Fujita's concern with global power, financial crisis, banking, debt crisis, urban insecurity, and economic inequalities is a great step towards demonstrating the multidimensionality and dynamism of urban crisis (2013, 1). By conceptualizing cities as embodiments of crisis and its aftermath, both due to spatial density and hosting protest movements (Ibid., 3), Fujita exposes the "lack of the crisis perspective in contemporary urban theories" (Ibid., 5). This concern with urban crises is also at the core of the European Programme for Sustainable Urban Development's 2010 report, which discusses the impact and responses to the 2009 and 2010 economic and financial crisis in various cities across Europe. These cities are Dublin in Ireland, Gijón in Spain, Jyväskylä in Finland, Malmö in Sweden, Newcastle in the United Kingdom, Rotterdam in the Netherlands, Tallinn in Estonia, Turin in Italy, and Veria in Greece. Arguing that "cities are on the front line when it comes to dealing with the real impacts of the crisis on people, business and places" (European Programme for Sustainable Urban Development 2010, 6), the report motivates enhanced conversations between cities, urban policymakers, and researchers. This report resonates with similar studies, such as Manuel Perlo

Cohen's 2011 study of the impact and responses to the 2008 global economic crisis within urban locales. Both studies commonly share the adoption of economic crisis as the major indicator of crisis in cities. On one hand, Fujita's scope of theorizing urban crisis is limited to Keynesian Crisis Theory's concern with the stability of the capitalist economy (2013, 7), and hence does not factor in other aspects of urban life that fall outside economic stability, but which typifies urban crisis today. On the other hand, the European Programme for Sustainable Urban Development locates crisis within the crumbling economic capacities of various European cities. However, crisis urbanism – this book's main concern – requires a re-theorization that accounts for human perspectives and actions in conjuring new urban practices. To this end, I start to consider the importance of other dimensions of urban crisis – taking them as starting points to larger urban configurations that precipitate, as inevitable aftermath, emergent forms of urbanism.

Another approach to discussing urban crises focuses on the typology of humanitarian crises, such as natural disasters, population displacement to cities, and conflict in cities (Fabre 2017, 2–3). These include natural disasters such as the 2010 earthquake in Port Au Prince, Haiti; the violent conflicts of Bangui, Central African Republic; the refugee crisis of Gaziantep, Turkey and Mafraq, Jordan; and the typhoon destruction of Bogo, Guiuan and Tacloban, Philippines (Pupulin, Gallet, and Decorte 2016). Although human vulnerability is central to the way urban crises are framed and discussed, a useful approach could be to abstract urban crisis as a manifold situation, emphasizing the need to understand "urban crisis contexts" by the international humanitarian community (Grindsted and Piquard 2009, 13). It is from such abstraction that we can theorize forms of resistances and coping mechanisms that rely on "new social behaviours, new social links, and new structures, values, and beliefs… a full re-creation, a re-adaptation to a new context" (Ibid., 17). Implicitly, urban crises result in protracted urban practices, which in turn generate distinct forms of urbanism.

Building on these ideas, this book proceeds on two premises. First, that urban crises, being multidimensional, are not fleeting happenstances, but accrue prolonged urban behaviours. In the end, any theory that may work would localize attention to the contextual happenstances both in terms of locale and long-term temporality, in order to effectively grasp these emergent urban practices. Second, that African city crises are not merely economic or political but are also entangled with postcolonial urban histories and their aftermaths. The resulting urban practices precipitated by these crises – hence crisis urbanism – are an indispensable resource in reformulating "the nature and potentialities of the urban" in Africa (Pieterse 2008, 109).

Crisis urbanism is thus used here as a "conceptual model of the city from the perspective of the slum … [rather than] from the perspective of the formal, concrete-and-steel city, as is normally done" (Ibid.). It designates the practices of adapting to the risky flows of contemporary African city life with its spectrum of difficulties. It is not a condition of urban spaces but a characteristic of ongoing or *in situ* practices of placemaking in a context of dynamic, vulnerable urban circumstances. In postcolonial African cities, crisis urbanism denotes the unconventional responses to urban exertions, undertaken either individually or collectively, and which splinter from or challenge mainstream urban practices. These splinter responses, it is argued, have become the main characteristic of urban life in almost all postcolonial African cities today. In fact, they are the very force that shapes how these cities function amidst the pressures of increasing human populations and global forces, dynamics made most evident in spaces of urban informality. In postcolonial African cities, within these informal spaces, one finds *sites* and *moments* of "imbuing with meaning" (Mbembe 2001, 7). These spaces are characterized by "people's relentless determination to negotiate conditions of turbulence and to introduce order and predictability into their lives" (Mbembe and Nuttall 2004, 349). The meaning of postcolonial African urbanism pursued here thus springs from these energizing moments of reflection about the history of human enterprise in these cities. The question is, however, whether African cities were in the past any more formal (in precolonial times? during colonialism? after independence?) than they are today? And what would this historical reality mean for this book's undertaking of a critique of Africa's informal urbanisms? To locate the ahistorical nature of this book's arguments within the continent's history of urbanism (inescapably implicit in the idea of "the postcolonial" central to the book), two considerations are important.

First, that an "ahistorical" reading of African cities would show crisis as a critical perspective of Africa's contemporary urbanism. Almost all existing literature on African cities cite crises in various forms: social, economic, political, spiritual, spatial, behavioural, governance, and so on. Yet, this is mostly the work of global north scholars, who underscore crises as inbuilt into Africa's urban form and do not explore its function across a wide scale. Even if it did, the history leading to such crises – for instance, the structural adjustment programmes that devastated African cities – does not refute the reality of crises or its imaginary and representation. Films like Abderrahmane Sissako's *Bamako* (2006), which offers a semiotic connection between Bamako's economic crisis and its colonial history, still dwells on the ahistorical everyday crisis to fully express its magnitude and impact in the form of human crisis in the postcolonial city. Second, in the same spirit of an ahistorical reading of African cities, the issue of informality may be focused on the present time. Thus, whereas African cities have been rapidly urbanizing for decades, the present moment offers greater insight into what these cities have become. The book thus seeks such an ahistorical reading of African cities as a contribution to the emergent perspectives on global north-south intellectual theories. In this sense, the book's investment in this "ahistoricity" of both African urban theories and African urban imaginaries may be read in the spirit of Comaroff and Comaroff (2014).

How Africa's urban dwellers work with incessant crises – customizing their responses to multiple difficulties to allow, even guarantee, survival – is a site offering novel ways of interpreting postcolonial urbanism. What incidental aspects of African cities are implied by these emerging forms of urbanism that enable sustained response to the constant crisis? Is it the familiarity of dislocation and dispossession, an excess of human ambition and capacities, or opportunistic manoeuvres characteristic of neocolonial capitalist mentalities? Considering that African cities are unique – with local situations that are not exactly replicated in other African cities – and that these pragmatic issues index local forms of crisis and the resulting urban practices that have become possible in response to these scenarios, crisis urbanism is thus conceptualized as a theory and method of interrogating such crises without over-generalizing. I see this as a strength, allowing this book to offer a diversity of urban perspectives about Africa, rather than a singular lens treating all African cities as if they were the same. Accordingly, the resultant inflow of research in this direction permits elastic discussions of the African city, not only from an anthropological or sociological or engineering perspective, but also from the standpoint of urban imaginaries. Many urban Africa-oriented filmmakers have used the city as their setting and its everyday survival as the raw material for their narratives. This relationship between urban crisis and urban cinema in Africa is historically grounded, favoured by the concentration of filmmaking technology and professionals in cities as well as the abundance of everyday experiences of crisis.

Accordingly, this book explores the cinema representations of the various forms of crisis urbanism across six African cities as an imaginary of an ongoing, even normalized, mode of urban life in Africa. Consequently, the book's reliance on film narratives and visual representations of postcolonial African cities posits creative imaginaries of cities as useful resources for theorizing urban Africa. To reiterate, the objective here is to illustrate how urban films may help support new research on postcolonial Africa crisis urbanism, and to do so in the context of the concepts of nonentity, laissez faire economic practices, urban commons, superficies of expendable citizenship, latencies of revolt, and experiences of fortification. The book's unique contribution to the wider discourse on cities and crisis is that it discusses crisis urbanism as the primary form of urbanism in postcolonial Africa. The book addresses two questions: how do postmillennial urban Africa films narrate crisis urbanism? And what discussions are possible about the resulting urban practices precipitated by these conditions as enablers of postcolonial Africa's urban theory? The responses sought here derive from the growing interest in the intersection between film narratives and urban studies. This book addresses Africa's postcolonial urbanism from the standpoint that a theorization of Africa's postcolonial urbanism through urban film narratives is both possible and overdue.

Postmillennial films and the African city

As a product of industrialization, film has tended to use cities to support its production needs (technology, crew, narrative setting, and stories) and marketing (exhibition and distribution). In urban Africa, the turn of the millennium saw a convergence of two developments: rapid urbanization and rapid proliferation of digital video technologies. The effects were equally entangled. Many rural populations relocating to cities generated new socio-economic tensions: unemployment and lack of opportunities meant that many of these city dwellers embraced informality as the mainstay model of urbanism. At the same time, with video cameras, video compact disks, and digital video disks becoming easily available, filmmaking industries sprouted, rapidly producing thousands of films – many of which used the city as a major setting, and diverse aspects of urban life as the main narrative material. These two occurrences have produced a recursive trend whereby urban pressures multiply with increasing populations and diminishing resources, and new urban narratives emerge from rapidly evolving survival mechanisms. Such urban narratives have generated huge impetus for many African and international filmmakers interested in telling postcolonial stories about African cities.

Alongside these developments, there has been a growing academic interest in interdisciplinary studies of cities through cultural arts, literature, and media (AlSayyad 2006; Lindner 2006; McQuire 2008; Stevenson 2003). There has also been sustained research on theorizing cities through films (Barber 2002; Brundson 2012; Bruno 2002; Clarke 1997; Donald 1999; Mazierska and Rascaroli 2002; Mennel 2008; Morcillo, Hanesworth, and Marchena 2015; Shiel and Fitzmaurice 2001; Silver and Ursini 2005). In Africa, the field has already attracted scholars studying African urbanism through literature and film (Dovey 2009; Frassinelli 2015; Khan 2016; Kruger 2006; Mututa 2019a, 2019b, 2019c, 2020; Nuttall 2008, 2009; West-Pavlov 2014). These discussions attest to the increasing significance of studying cities through films. It is for this reason that this book has turned to postmillennial African cinema, the fastest-growing urban archive of Africa's postcolonial urbanism, as a potential resource for studying and theorizing the continent's cities.

Responding to the scarcity of studies exploring the representation of changing patterns of African urbanism in cinema, this book focuses on the study of African urbanism through films. The book's quest is to discuss the relation between increasing urban populations in Africa, rapid adoption of digital video technologies, and the increase of narratives concerning the continent's urban crises. Using video narratives as archives of contemporary urbanism, *Crisis Urbanism and Postcolonial African Cities in Postmillennial Cinema* explores the role that postmillennial African films may play in theorizing Africa's postcolonial urbanism. It straddles two disciplines: African urban studies and African cinema. Engaging with film narratives on one hand and existing urban theories on the other, the book seeks to orient ongoing scholarship on African urbanism towards urban film narratives as a useful resource for reading and theorizing postcolonial African urbanism. The book further recognizes that rampant crises at individual and communal levels in African cities are more the norm than exception. As African cities evolve into chaos, urban residents have had to adopt radical manoeuvres to survive. Becoming a nobody in post-1994 Johannesburg's townships, private oppressions and resistances in Cairo, postwar superficies in Monrovia, urban outliers in postwar Luanda, wrangles over city commons in Nairobi, and dire economic conditions in Kinshasa are all exemplary of mainstay urban crises. "Crisis urbanism," as I call this rampant form of problem-filled urban life in Africa today, demands fresh insights to grasp its workings. Beyond the city of concrete and steel lies a city of experiences, routines, decisions, and random events. This city is spontaneous, disjointed from place to place, and a consequence of the crises that characterize it. This urbanism, formed by the crisis that it seeks to solve, offers a new tangent for studying urban Africa today.

The method of film analysis used here is a close reading of the films, abstracting the minutiae of crisis urbanism in postcolonial Africa as tools to work within the context of current debates on southern urbanism theory. A crime here, an idle group there, a conversation, a gesture, a building, a road-side vendor, a police officer, a relationship, a garbage heap, a momentary glimpse, and so on are all raw materials for expressing the human condition amidst various urban pressures. The book hypothesizes that because these film narratives rely on urban formations, networks, relations, events, and experiences to show the various facets of urban crises across different African cities, such elements contain crucial cues about Africa's postcolonial urbanism. By focusing on crisis as the dominant paradigm of Africa's postcolonial urbanism, *Crisis Urbanism and Postcolonial African Cities in Postmillennial Cinema* takes the debate away from stereotypical theorizations of African cities and instead seeks an evocative approach to reading cities. The book is fully aware that southern urbanism is no longer sufficiently conceptualizable as an outcome of colonial geopolitical or economic posterity. Rather, it is very much about the present – and its implications for the urban future – and, cognizant of earlier and ongoing labours to provide a southern urbanism theory, it offers crisis urbanism as a pragmatic theory with which to study African cities today. Also, fully aware of the ongoing pursuit of alternative perspectives on Africa's urban theory (Myers 2011), this book approaches African urbanism as a "problem-space" (Roy 2011, 307) that can be understood on its own terms, through both empirical work and cultural imaginaries, with this book pursuing the latter.

The book

Crisis Urbanism and Postcolonial African Cities in Postmillennial Cinema offers a reasoned critique of the various pragmatic crises that have come to beleaguer, and incidentally characterize, postcolonial African cities. These include practices of nonentity in Johannesburg's post-apartheid townships, the callousness of informal economies in Kinshasa, power in the context of urban commons in Nairobi, violence as hustle in postwar Monrovia, resisting governance of citizenship in Cairo, and the constraints of fortified life in postwar Luanda. By covering these six cities, the book aims to engage the various debates on Africa's postcolonial urbanism, while establishing a synthesis between urban films and urban theory. A summary of the book's key concerns in the various chapters follows below.

Chapter 1 contextualizes the idea of crisis in Africa's postcolonial urbanism. It frames crisis urbanism within recent debates on the continent's urban life, drawing attention to the need for an interdiscipnary analysis of cities. Further, it unpacks the convergence between the anthropological study of Africa's urbanism and postmillennial cinema. The introduction part argues that an interdisciplinary analysis of urban crises that juxtaposes urban theory with cinema theory, an erstwhile unpopular norm in the continent's academia, may indeed offer a critical entry point into studying qualitative aspects of postcolonial urbanism in Africa.

Chapter 2 is anchored in the lack of urban identity among the residents of Johannesburg's townships at the end of apartheid. Terming this condition "nonentity," I argue that metaphors of the township as an "unknowable space" and its residents as the "urban unknowns" have usurped the apartheid metaphors of belligerence and oppression. Indeed, the gentrification, improvements in infrastructural development, and access to utilities and services that have occurred in townships since 1994 have alleviated overt apartheid metaphors of social, economic, and political bias. Current debates on the township are focusing on these emergent, latent urban challenges (Duminy, Parnell, and Brown-Luthango 2020). By looking beyond the township's past symbolism of apartheid distress, this chapter seeks new meanings of South Africa's post-apartheid urbanism through a critique of the representation of South Africa's townships in Garvin Hood's *Tsotsi* (2005). It argues that the township has discarded its apartheid-era imagery of racial binary without acquiring an alternative label to account for the ways in which it has changed. Designating this borderline status as "nonentity," the chapter hypothesizes that the images of the township afford a chance to seize visual filmic codes and theorize an evolving imaginary of the township in contemporary South Africa. The chapter's key argument is twofold: that the urban mise-en-scène is a signifier of an unintelligible township and that the township is a modular narrative space that expresses the recursion of marginalization. Both aspects index nonentity – conveyed through *Tsotsi's* formal film visual style – as a paradigmatic label of the new South African township dweller, designating the kind of offsite life that is possible in post-apartheid times.

Like many postcolonial African cities, Kinshasa has evolved towards informality. This is most evident in the proliferation of informal street economies. In the city's periphery, these activities feed into the perception of popular practices of informal survival in postcolonial urban Africa as not "constrained by its material conditions" but instead offering "something no legal neighbourhood could: freedom" (Neuwirth 2005, 5). Street vending anchored in individual enterprise offers a template to speculate on futile urban survival, and hence acts as a provocateur of the shadowy forms of the urban that spring up precisely in hotspots of urban crises. Chapter 3 discusses these economies of dystopia, as found in postcolonial Kinshasa. Labelling the resulting shadowy city life – a space where the established procedures guiding urban life in many formal spaces are hardly applicable

(Simone 2007, 84) – as *"laissez-faire* urbanism," the chapter underscores the increasing centrality of unconventional economic practices in characterizing Kinshasa's postcolonial urban trends. Through a close reading of Djo Tunda wa Munga's representation of Kinshasa in *Viva Riva!* (2010), this chapter illustrates the way individual-based survivalist entrepreneurship, labelled *"laissez faire* economic practices," becomes a placeholder for Kinois' guarded mode of urban survival. The chapter taps into existing local terminologies circulating in Kinshasa – *Khadaffi* and *la sapeur* – to illustrate how these characters' economic practices in the face of ever-present danger represent ingenious responses to urban economic crises.

In many aspects, Nairobi city has become a conglomerate of informal habitation. With the city's affluent squeezed in pockets of high-end suburbs surrounded by miles of informal settlements, the city in its current form is mainly informal. Although this structure is a rip-off of urban patterns seen around the continent, it nevertheless has major implications for how the city's space may be theorized in the present. The incapacity to share the city as a common space – a failure that sprouts from contestation over real estate as a scarce urban resource – also produces the rationale that certain urban spaces would be sustainably developed and others, relegated as "trash," would attract insufficient capital investment. This has produced a skyline of dilapidated material structures and modern skyscrapers, both forming a visible frontier between the two cities. In this context, garbage acquires the metonymic value of designating the boundary between the formal and the informal and the legal and the illegal. The operationalization of trash goes beyond the scattering of consumer product wraps and rotting perishable commodities, leaking sewers and chemical effluents, and the various shades of rot and pollution oozing in some parts of the city. Trash here is an allegory of the everyday life of Nairobi residents, who, living lives infiltrated by garbage, inhabit a "trashed city," the only possible site where they can gain inclusion to the city's dwindling spaces and resources. Here, inhabiting trashed spaces expresses a rapidly evolving form of urbanism where trash (and emblematic spaces such as downtown) becomes the only option to claim portions of the city for those with insufficient economic means. Based on this indexicality of trash and the role it plays in designating ways of communing with Nairobi city, Chapter 4 looks at trash's indexicality of urban power tussles, examining how Gitonga's *Nairobi Half Life* (2012) uses Nairobi's garbage sites to motivate a debate on the urban commons. "Occupation of trash," it is argued, designates a form of crisis urbanism that experiments with tactics around sharing the city as a common space.

Chapter 5 discusses *rarray* citizens as the poster image of the crisis of postwar hustle in Monrovia. Urbanism in Monrovia, as in many other parts of Africa, is characterized by persistent "postcolonial" crisis. The reference to postcolonialism is, of course, provocative when referring to Liberia, a country where no real colonization happened. In the absence of colonial history, Monrovia presents a unique case of operationalizing crisis as a way of urban life. Monrovia's most notable crisis involved violent seizure by warring factions during years of civil war. In retrospect, the imprint of this history is strongly present not just in the imaginaries of the wartime city but also in the everyday life in postwar Monrovia. Various films set in Monrovia show the city as a watershed for a residual sense of disorientation and displacement, both in physical terms (a significant aspect of the wartime period), and also in social, economic, and political terms. This is most evident in the characterization of disenfranchised urban residents who, despite occupying physical spaces in the city, exist superfluously with little connection to the city's social and economic flows. This chapter discusses the representation of urban displacement and its operative purpose in the character of the rarray – a postwar urban character displaced from social, economic, and political affiliation – as an authentic urban construct. Through a close reading of the rarray in Andrew Niccol's *Lord of War* (2005) and Jean-Stephane Sauvaire's *Johnny Mad Dog* (2008), the chapter explores the rarray as emblematic of west and central Africa's postwar paradigm of emergent urbanism. The chapter further discusses two key characterizations of rarray urbanism, specific to Monrovia cinema; first, in the context of a wartime necropolis; and second, in the context of postwar urban fractures. The key argument is that the rarray character is paradigmatic of existential crisis in the city and hence offers an aperture into the superficies of postwar crisis urbanism in Monrovia.

When the so-called Arab Spring finally landed in Egypt, it not only marked new approaches to negotiating and shaping governance, it also marked a unique approach to responding to various pressures of citizenship. One obvious realization was that, behind the outlook of civility and obedience to governance, a seething rage had accrued. Starting from this insight, Chapter 6 discusses Cairo's crisis citizenship as a venerable tug-of-war between latent oppressive manoeuvres and covert rebellion, a scenario that has produced an urbanism of revolt in cities like Cairo, where the effects of this struggle were most evident. Focusing on the representations of this emergent urbanism in Khaled Marie's *Asal Aswad* (*Molasses*) (2010), this chapter builds a theoretical connection between the overt street protests that erupted in the city in 2011 and the covert disenfranchisement that existed long before then. Post-2011 films and media reports emphasising the explicit crisis form the bulk of this theorization. The chapter's main argument is that these urban film narratives convey an urban crisis in the form of latent oppression and resistance in everyday life in Cairo. This covert crisis, more than its climax as the Egyptian revolution, conveys the underlying urban crisis, and thus provides a useful perspective on the meaning of urban citizenship in postcolonial Cairo. The chapter works across a range of symbols such as immigration, passports, transport, freedom, and state services as frontiers of this crisis.

Chapter 7 explores the urbanism of fortification in postwar Luanda. The modes of inhabiting city space in postcolonial Luanda is historically eclectic. Even without extreme and obvious spatial rules that dictate modes of urban tenancy, covert demarcations still exist in postwar Luanda. Such delineations ensue from the ever-present existential barriers to accessing the infrastructure necessary for favourable urban survival. This imaginary permeates many postmillennial films about this city. Postwar Luanda materializes in José Augusto Octávio Gamboa dos Passos' (Zézé Gamboa) *O Herói (The Hero)* (2004) as a city fortified by intrinsic yet unseen barriers. The survival of the residents of postcolonial Luanda appears, at a glance, to be constituted along rigid politicized social and economic structures. Thus, understanding popular cultural imaginaries of Luanda as a restrictive city implicates the underhanded workings of the postcolonial welfare state and the repercussions of its soft power (Watson 2002, 49). Without delving into the political economies of such relations between the state and postcolonial urbanism in Angola, the question of citizenship rights, or the intersection of politics of postwar urbanism and political manoeuvres, this chapter explores the representations of fortification in postwar Luanda in Gamboa's *O Herói*. The chapter argues that the protagonist's confinement in wretched urban life – evident in his futile pursuit of healthcare, job opportunities, and better living conditions – is emblematic of the project of exclusive postcolonial urbanism in Luanda. Using what I term "cantonment" and "outlier urbanism" as a conceptual lens, the chapter critiques *O Herói's* representation of Luanda's urban form and the protagonist's experiences in various urban sites as constituting a useful perspective on Luanda's postwar urbanism. That the film's characters' oscillation towards a better life is juxtaposed with a sense of despair, it is argued, portrays an elaborate scheme of inescapable alienation within the city. The theorization that follows is twofold. First, that urban characters coping with inescapable incapacitation in their pursuit of postcolonial inclusion exemplify the likeness to postwar cantonment; and, second, that such urban restriction produces peculiar urban practices, subsequently theorized as "outlier urbanism." Both aspects are paradigmatic of Luanda's postwar crisis urbanism.

Certainly, the overarching theoretical underpinning of the book is that various crises in urban Africa have not only shaped life in postcolonial African cities but also stimulated and supported the sprouting and flourishing of new forms of urbanism. Chapter 8 summarizes the key arguments raised in the book and suggests further debates and possible future trajectories of crisis urbanism as an ongoing paradigm of Africa's postcolonial urbanism.

Bibliography

Alcaly, Roger E., and David Mermelstein. 2006. *The Fiscal Crisis of American Cities*. New York and Toronto: Vintage.

AlSayyad, Nezar. 2006. *Cinematic Urbanism: A History of the Modern from Reel to Real*. New York: Routledge.

Barber, Stephen. 2002. *Projected Cities: Cinema and Urban Space*. London: Reaktion.

Bhabha, Homi K. 1994. *The Location of Culture*. London and New York: Routledge.

Brundson, Charlotte. 2012. "The Attractions of the Cinematic City." *Screen* 53 (3) Autumn: 209–227.

Bruno, Giuliana. 2002. "Site-Seeing: The Cine City." In *Atlas of Emotion: Journeys in Art, Architecture, and Film*, edited by Giuliana Bruno, 15–54. London: Verso.

Castells, Manuel. 1977. *The Urban Question: A Marxist Approach*. London: Edward Arnold.

Çınar, Alev, and Thomas Bender. 2007. *Urban Imaginaries: Locating the Modern City*. Minneapolis and London: University of Minnesota Press.

Clarke, David B., ed. 1997. *The Cinematic City*. London and New York: Routledge.

Cohen, Manuel Perlo. 2011. *Cities in Times of Crisis: The Response of Local Governments in Light of the Global Economic Crisis: The Role of the Formation of Human Capital, Urban Innovation and Strategic Planning*. Working Paper 2011–01, Berkeley, CA: Institute of Urban and Regional Development University of California - Berkeley.

Comaroff, Jean, and John L. Comaroff. 2014. *Theory from the South: How Euro-America Is Evolving Toward Africa*. Stellenbosch: SUN MeDIA.

De Boeck, Filip, and Marie-Francoise Plissart. 2004. *Kinshasa: Tales of the Invisible City*. Leuven: Leuven University Press.

De Satge, Richard, and Vanessa Watson. 2018. *Urban Planning in the Global South: Conflicting Rationalities in Contested Urban Space*. London and New York: Palgrave Macmillan.

Dirlik, Arif. 2007. "Global South: Predicament and Promise." *The Global South* 1 (1): 12–23.

Donald, James. 1999. "Light in Dark Spaces: Cinema and the City." In *Imagining the Modern City*, edited by James Donald, 63–94. Minneapolis: University of Minnesota Press.

Dovey, Lindiwe. 2009. *African Film Adapting Violence to the Screen*. New York: Columbia University Press.

Duck, Leigh Anne, ed. 2015. *The Global South*. Bloomington: Indiana University Press.

Duminy, James, Susan Parnell, and Mercy Brown-Luthango. 2020. *Supporting City Futures: The Cities Support Programme and the Urban Challenge in South Africa*. Cape Town: African Centre for Cities.

Eckardt, Frank. 2015. "City and Crisis: Learning from Urban Theory." In *City of Crisis: The Multiple Contestation of Southern European Cities*, edited by Frank Eckardt and Javier Ruiz Sánchez, 11–30. Bielefeld: Transcript Verlag.

European Programme for Sustainable Urban Development. 2010. *URBACT Cities Facing the Crisis: Impact and Responses*. URBACT November 2010, Saint-Denis La Plaine: URBACT Secretariat.

Fabre, Cyprien. 2017. *Urban Crisis: World Humanitarian Summit - Putting Policy into Practice*. Humanitarian Aid and Civil protection (ECHO), André-Pascal: Development Co-operation Directorate, OECD.

Foster, Jeremy. 2009. "From Socio-nature to Spectral Presence: Re-imagining the Once and Future Landscape of Johannesburg." *Safundi: The Journal of South African and American Studies* 10 (2): 175–213.

Frassinelli, Pier Paolo. 2015. "Heading South: Theory, Viva Riva! and District 9." *Critical Arts* 29 (3): 293–309.

Fujita, Kuniko, ed. 2013. *Cities and Crisis: New Critical Urban Theory*. London: SAGE Publications.

Gastrow, Claudia. 2017. "Cement Citizens: Housing, Demolition and Political Belonging in Luanda, Angola." *Citizenship Studies* 21 (2): 224–239.

Gastrow, Vanya. 2018. *Problematizing the Foreign Shop: Justifications for Restricting the Migrant Spaza Sector in South Africa*. Samp Migration Policy Series No. 80. Cape Town: Southern African Migration Programme.

Grindsted, Annette, and Bridgitte Piquard. 2009. "Cities and Crises." In *Cities and Crises*, edited by Dennis Day, Annette Grindsted, Bridgitte Piquard, and David Zammit, 13–24. Bilbao: Univesity of Deusto.

Harding, Alan, and Talja Blokland. 2014. *Urban Theory: A Critical Introduction to Power, Cities and Urbanism in the 21st Century*. London: Sage Publications.

Khan, Khatija. 2016. "Representations of Crime, Power and Social Decay in the South African Post-Colony in the Film Gangster's Paradise: Jerusalema (2008)." *Communicatio* 42 (2): 210–220.

Koonings, Kees, and Dirk Kruijt. 2009. *Megacities: The Politics of Urban Exclusion and Violence in the Global South*. London and New York: Zed Books.

Kruger, Loren. 2006. "Filming the Edgy City: Cinematic Narrative and Urban Form in Postapartheid Johannesburg." *Research in African Literatures* 37 (2): 141–163.

Léfebvre, Henri. 1991. *The Production of Space*. Translated by Donald Nicholson-Smith. Oxford: Blackwell.

Léfebvre, Henri. 1996. *Writings on Cities*. Edited by Eleonore Kofman and Elizabeth Lebas. Translated by Eleonore Kofman and Elizabeth Lebas. Oxford: Blackwell Publishers.

Lindner, Christoph, ed. 2006. *Urban Space and Cityscapes: Perspectives from Modern and Contemporary Culture*. London and New York: Routledge.

Locatelli, Francesca, and Paul Nugent. 2009. *African Cities: Competing Claims on Urban Spaces*. Leiden and Boston: Brill.

Madani-Pour, Ali. 1995. "Reading the City." In *Managing Cities: The New Urban Context*, edited by Patsy Healey, Stuart Cameron, Simin Davoudi, Stephen Graham, and Ali Madani-Pour, 21–26. Chichester, West Sussex: John Wiley & Sons Ltd.

Mazierska, Ewa, and Laura Rascaroli. 2002. *From Moscow to Madrid: Postmodern Cities, European Cinema*. London: Bloomsbury Publishing.

Mbembe, Achille. 2001. *On the Postcolony*. Berkeley/Los Angeles/London: University of California Press.

Mbembe, Achille. 2008. "Aesthetics of Superfluity." In *The Elusive Metropolis*, edited by Sarah Nuttall and Achille Mbembe, 37–67. Durham and London: Duke University Press.

Mbembe, Achille, and Sarah Nuttall. 2004. "Writing the World from an African Metropolis." *Public Culture* 16 (3): 347–372.

Mbembe, Achille, and Sarah Nuttall. 2008. "Introduction: Afropolis." In *The Elusive Metropolis*, edited by Sarah Nuttall and Achille Mbembe, 1–36. Durham and London: Duke University Press.

McQuire, Scott. 2008. *The Media City: Media, Architecture and Urban Space*. Los Angeles: Sage Publications.

Mennel, Barbara. 2008. *Cities and Cinema*. London and New York: Routledge.

Morcillo, Marta García, Pauline Hanesworth, and Óscar Lapeña Marchena. 2015. *Imagining Ancient Cities in Film: From Babylon to Cinecittà*. New York and London: Routledge.

Murray, Martin J. 2008. *Taming the Disorderly City: The Spatial Landscape of Johannesburg After Apartheid*. Ithaca, NY: Cornell University Press.

Murray, Martin J. 2011. *City of Extremes: The Spatial Politics of Johannesburg*. Durham and London: Duke University Press.

Mututa, Addamms Songe. 2019a. "Johannesburg in Transition: Representing Street Encounters as Racial Registers in Clint Eastwood's Invictus." *Critical Arts South-North Cultural and Media Studies* 33 (2): 1–13.

Mututa, Addamms Songe. 2019b. "Vertical Heterotopias: Territories and Power Hierarchies in Tosh Gitonga's Nairobi Half Life." In *Narratives of Place in Literature and Film*, edited by Steven Allen and Kirsten Møllegaard, 158–170. New York and London: Routledge.

Mututa, Addamms Songe. 2019c. "The Casebre on the Sand: Reflections on Luanda's Excepted Citizenship Through the Cinematography of Maria João Ganga's Na Cidade Vazia (2004)." *Journal of African Cinemas* 11 (3): 95–111.

Mututa, Addamms Songe. 2020. "Customizing Post-apartheid Johannesburg: The Dialectic of Errancy in Ralph Ziman's Jerusalema." *Safundi: The Journal of South African and American Studies* 21 (2): 206–223.

Myers, Garth. 2011. *African Cities: Alternative Visions of Urban Theory and Practice*. London and New York: Zed Books.

Nadarajah, Manickam, and Ann Tomoko Yamamoto. 2007. *Urban Crisis: Culture and the Sustainability of Cities*. Tokyo, New York and Paris: United Nations University Press.

Nas, Peter J.M., Marlies de Groot, and Michelle Schut. 2011. "Introduction: Variety of Symbols." In *Cities Full of Symbols: A Theory of Urban Space and Culture*, edited by Peter J.M. Nas, 7–26. Amsterdam: P.J.M. Nas/Leiden University Press.

Neuwirth, Robert. 2005. *Shadow Cities: A Billion Squatters, a New Urban World*. London and New York: Routledge.

Nuttall, Sarah. 2008. "Literary City." In *Johannesburg: The Elusive Metropolis*, edited by Sarah Nuttall and Achille Mbembe, 195–218. Durham and London: Duke Unveirsity Press.

Nuttall, Sarah. 2009. *Entanglement: Literary and Cultural Reflections on Post-apartheid*. Johannesburg: Wits University Press.

Parnell, Susan, and Sophie Oldfield. 2014. *The Routledge Handbook on Cities of the Global South*. London and New York: Routledge.

Parnell, Susan, and Jennifer Robinson. 2012. "(Re)theorizing Cities from the Global South: Looking Beyond Neoliberalism." *Urban Geography* 33 (4): 593–617.

Pieterse, Edgar. 2008. *City Futures: Confronting the Crisis of Urban Development*. Lansdowne: UCT Press.

Pieterse, Edgar. 2011. "Grasping the Unknowable: Coming to Grips with African Urbanisms." *Social Dynamics: A Journal of African Studies* 37 (1): 5–23.

Pieterse, Edgar. 2014. "*Epistemological Practices of Southern Urbanism.*" *Draft Paper to be presented at the ACC Academic Seminar*. University of Cape Town.

Pupulin, Luca, Bertrand Gallet, and Filiep Decorte. 2016. *Consultations on Humanitarian Responses in Urban Areas: Perspectives from Cities in Crisis*. World Humanitarian Summit, May 2016, Geneva and Barcelona: Global Alliance for Urban Crises.

Roy, Ananya. 2011. "Slumdog Cities: Rethinking Subaltern Urbanism." *International Journal of Urban and Regional Research* 35 (2): 223–238.

Shiel, Mark, and Tony Fitzmaurice. 2001. *Cinema and the City: Film and Urban Societies an a Global Context*. Oxford: Blackwell Publishers.

Silver, Alain, and James Ursini. 2005. *L.A. Noir: The City as a Character*. Santa Monica: Santa Monica Press.

Simone, AbdouMaliq. 2007. "Assembling Douala: Imagining Forms of Urban Sociality." In *Urban Imaginaries: Locating the Modern City*, edited by Alev Çinar and Thomas Bender, 79–99. Minneapolis: University of Minnesota Press.

Simone, AbdouMaliq. 2008. "Some Reflections on Making Popular Culture in Urban Africa." *African Studies Review* 51: 75–89.

Southall, Aidan. 1998. *The City in Time and Space*. Cambridge: Cambridge University Press.

Soyata, Tolga, Hadi Habibzadeh, Chinwe Ekenna, Brian Nussbaum, and Jose Lozanod. 2019. "Smart city in crisis: Technology and policy concerns." *Sustainable Cities and Society* 50 (October 2019): 1–15. DOI: 101566

Stevenson, Deborah. 2003. *Cities and Urban Cultures*. Maidenhead and Philadelphia: Open University Press.

Watson, Sophie. 2002. "The Public City." In *Understanding the City: Contemporary and Future Perspectives*, edited by John Eade and Christopher Mele, 49–65. Oxford: Blackwell Publishers.

West-Pavlov, Russell. 2014. "Inside Out – The New Literary Geographies of the Post-Apartheid City in Mpe's and Vladislavić's Johannesburg Writing." *Journal of Southern African Studies* 40 (1): 7–19.

Wickham, Chris. 2015. *Medieval Rome: Stability and Crisis of a City, 900–1150*. Oxford: Oxford University Press.

Filmography

Asal Aswad (Molasses). Dir. Khaled Marie, Egypt, 2010.

Bamako. Dir. Abderrahmane Sissako, Mali, 2006.

Johnny Mad Dog. Dir. Stephane Sauvaire, France, 2008.

Lord of War. Dir. Andrew Niccol, USA, 2005.

Nairobi Half Life. Dir. Tosh Gitonga, Kenya, 2012.

O Herói (The Hero). Dir. Zézé Gamboa, Angola, 2004.

Tsotsi. Dir. Garvin Hood, South Africa, 2005.

Viva! Riva. Dir. Djo Tunda Wa Munga, The DRC, 2010.

2 Crisis of nonentity

"Unknowing" Johannesburg's post-apartheid townships

Introduction

One of the most contested aspects of black experiences of Johannesburg was the Pass, which ensured, among other things, that the identity of the black urban dweller was well-known and documented (Guelke 2005). However, in post-apartheid Johannesburg where the Pass is no longer used, the unknowability of the city's black residents has assumed new meanings beyond the initial signification of freedom from apartheid-era surveillance and influx control mechanisms. One of the inadvertent consequences of uncontrolled mass movement into the city and townships was the emergence of "urban unknowns" as a new category of urban dwellers. Building on recent debates on the township which are increasingly interested in emergent, latent urban challenges (Duminy, Parnell, and Brown-Luthango 2020), this chapter discusses the lack of urban identity among the residents of Johannesburg's townships at the end of apartheid. Terming it "nonentity," the chapter argues that metaphors of the township as an "unknowable space" and its residents as "urban unknowns" have usurped the apartheid metaphors of belligerence and oppression. Building on emergent aspects of post-apartheid township life – improved infrastructural development, gentrification, and access to utilities and services – and attentive to how these may have alleviated overt apartheid metaphors of social, economic, and political bias, the chapter explores unknowability or lack of identity as one of the ways in which post-apartheid township residents suffer estrangement.

By looking beyond the township's past symbolism of apartheid distress, this chapter seeks new meanings of South Africa's post-apartheid urbanism through a critique of the representation of the township and its residents as unidentifiable in Garvin Hood's *Tsotsi* (2005). It argues that the township has discarded its apartheid-era imagery of racial binary, which relied on strict identities, without acquiring alternative labels to account for its change. Instead, the township has become a space where "formal or effective governance is absent. In concrete terms this means that there is no effective presence of state power and public institutions" (Koonings and Kruijt 2009, 1). Designating this status as nonentity, the chapter hypothesizes that the film's aesthetic representation of the township affords a chance to theorize an

DOI: 10.4324/9781003122098-2

evolving imaginary of the township in contemporary South Africa as a kind of offsite space in post-apartheid times. This aesthetic is conspicuously increasing in recent South African films.

The new urban South Africa in cinema

There are "particular moments in history that are defined by photographic, celluloid, or television images" (Simbao 2007, 52). This idea is based on Sam Nzima's iconic photograph of Hector Pieterson, Mbuyisa Makhubu, and Antionette Pieterson showing the fatal violence of the 1976 Soweto Uprisings in South Africa, which has since been recognized as the turning point of the apartheid struggle. However, it equally applies in the critique of the post-apartheid township imaginary in cinema, a medium that also relies on celluloid and digital technologies to capture specific moments of a rather dynamic urban life. Particularly in the new South Africa, where "the concept of the 'nation' is largely a fantastical or imaginative one," as Lindiwe Dovey (2009, 51) asserts, "art has a peculiarly privileged role in sustaining and questioning it." Present-day South African filmmakers, she adds, "are playing an important role in narrating ... nation into being" (Ibid.). Dovey cites, among other films, *Max and Mona* (2004), *Hijack Stories* (2000), *Tengers* (2007), *Conversations on a Sunday Afternoon* (2005), *Jump the Gun* (1997), *God Is African* (2001), *Cape of Good Hope* (2004), *Proteus* (2003), *Chikin Biznis – The Whole Story* (1999), and *Rape for Who I Am* (2005). These films question a range of post-1994 issues, including immigration, multiculturalism, disease, education, urban violence, and general urban dysphoria. Dovey (2009, 51) further argues that some films question "race, freedom, and violence in the 'new' South Africa." At the centre of this questioning is the image of a post-apartheid township, the most publicized structural symbol of South Africa's past, but now a monument that has been stripped of this symbolism and left without a replacement. If the township's past typified apartheid ideals, how do we make sense of its present and future, when official apartheid has been replaced with the rhetoric of a Rainbow Nation? What metaphors of South Africa's post-apartheid urbanism does a speculation into the future of the township allow?

These questions are the starting point for the discussions on nonentity as a post-apartheid township paradigm explored here. They frame South Africa's post-1994 townships already being imagined in films, such as *Wooden Camera* (2003) and *Max and Mona* (2004), and television series, like *Yizo Yizo*, *Gazlam*, and *Tsha Tsha*, which use post-1994 township space to mobilize myths about blackness (Ellapen 2007, 114). Similarly, *Zulu Love Letter* (2004), *Battle for Johannesburg* (2010), and *Miners Shot Down* (2014), which show the difficulties of blackness in post-apartheid urban South Africa, pay little attention to the apartheid past, instead questioning the present as a moment being experienced outside the frame of the expected urbanism. Equally, *District 9* (2009), described as an "intersection of temporalities that superimpose a dystopian view of the present – as opposed to the image of a possible future that is advancing upon us" (Frassinelli 2015, 301), imagines an invasive, unfamiliar future of the township for which residents are ill-prepared. We could add Clint Eastwood's *Invictus* (2009), which deals with the problems of post-apartheid racial harmony (Mututa 2019); and *Jerusalema: Gangsters' Paradise* (2008), which explicitly uses the proliferation of crime culture among township youth to comment on critical ideas of post-1994 urban incursions (Mututa 2020). The main characters, Nazareth (Jeffrey Zekele), Lucky Kunene (Rapulana Seiphemo), and Zakes Mbolelo (Ronnie Nyakale), tap into the generic identity of post-apartheid crime culture to engage with profound issues of urban malaise in a globalizing space rather than the historicity of the township. Through this film, we see a new template of the township taking shape: a township that is removed from the margins and detached from the physical space historically known as the township. The new embodied township is on the move, it can mingle with other races freely, it can celebrate, encourage, even receive support. But it is always there, an amorphous entity in every conceivable practice of urban life.

Jerusalema's story starts in 1994, with street scenes showing South Africans celebrating freedom, accompanied by Kunene's voiceover: "I had dreams. The dreams were the circulatory system of the black urban economy ... however free enterprise was never encouraged." At this point, the youthful Kunene and Zakes disembark from a moving train, fleeing a black security guard opposed to onboard hawking. When Lucky meets Nazareth, a former exiled freedom fighter turned street criminal, we learn of the emerging limbo surrounding township life:

LUCKY: We didn't fight the struggle so that we could become criminals.
NAZARETH: And I didn't fight apartheid to be poor either.

This dialogue reflects the cinematic concern with the township's "[d]iscourses of introspection and [self] critique" (Dovey 2009, 53), which does not end with a concrete framing of the new township, but rather in a denial of its past symbolism of black squalor. It is comparable to the practices of urban gentrification specific to post-1994 Johannesburg. In the previously whites-only residential zone of Hillbrow, one now finds flats commandeered by blacks inhabiting space in the model of the township, that is, with alcohol, prostitution, garbage, violence, and squalor (Mututa 2020). Here, hijacked urban spaces are quickly re-converted to slum-like conditions. Arguably, then, such urban perspectives mediated through film alert us to the problem of articulating the township's future without considering the textual meanings of its mundane practices and representations. This trend can be well understood as futuristic rather than historical.

The future of South Africa's townships

That "studies looking into the [city] future are rather few and tentative," argues Henri Léfebvre (1996, 211), is a "serious error." He continues: "work on the urban cannot limit itself merely to recording what has been produced;"

instead, urban theorists "must also look ahead and propose things" (Ibid.). Léfebvre speaks about Paris' rapidly changing socio-economic structures in the context of globalization. Yet, his attempt to propose Paris' future signposts the need to go beyond a city's history and account for its tentative future. For post-apartheid South African townships, this focus on the future has both great benefits and challenges.

AbdouMaliq Simone (2004, 3) terms such an undertaking a "daunting one in that it is difficult to ascertain with any precision just what kind of urban possibilities and futures are being made." In the case of post-apartheid South Africa, Sarah Nuttall (2006, 267) uses the term "emergent" to designate "a politics which addresses the potential, both latent and actively surfacing in South Africa … while taking cognizance of the ongoing ambivalences of the moment." Although there are many possibilities in attempting to heed Nuttall's call to pay attention to "new" or emerging interpretations of South African cities that disengage from the past to address the present and the future, it is equally plausible to fulfil this purpose by paying attention to the mundane. In his work on city futures in Africa, Edgar Pieterse (2008, 1) points out that "(n)othing about cities in the twenty-first century is insignificant; the stakes are always high in pinning down what cities are, in thinking about what to do with cities and in acting on/ in/through the city." He uses concepts such as "open-ended futures, or at least malleable futures" (Ibid., 5) and "truncated futures" (Ibid., 124) as piecemeal potentialities of urban South Africa's future theory. All these scholars agree that post-1994, the new urban Africa requires new theoretical framing that can account for disruptive – non-statistical, non-scientific, non-systematic – forms of future urbanism. This chapter joins this conversation by exploring nonentity as a non-scientific theory of South Africa's townships' crisis post-1994.

In the past, South Africa's townships indexed crises either in the form of political conflicts (Baines 2007) or racial segregation (Badcock 2002, 180). That Johannesburg's early slums – Orlando and Pimville (renamed South Western Townships, or SOWETO, in 1964) – were first inhabited by non-white South Africans evicted from Sophiatown in the 1950s, illustrates both crises. Loren Kruger (2013, xxiv) describes the township as a shanty peri-urban outpost for blacks, while Julian Brown (2016) and Baruch Hirson (1979) associate it with diminished rights to access and participate in social, political, and economic activities. Rogerson and Da Silva (1988, 255) term the townships as "'dormitories' or 'labour reservoirs' for the so-called 'White' cities". Here, blacks were schooled to be workers for white enterprises (Frueh 2003, 41). The township of the past was politically and racially imagined (Hart 1986). However, post-1994, the township exists within a political and racial framework that has undergone substantial revisions, including official reconstitution. There are no longer barbed wire fences or relocation and re-education camps and "what was once intended as an apotheosis of every tyrannical and authoritarian system that justified itself on the basis of racial differences, had been supplanted by a democracy led by a democratically elected black President" (Canepari 2011, 217). Canepari's inference resonates with other concerns that the historical symbolism of the township has become outdated (Kruger 2013, 97).

Various urban scholars have shown interest in pursuing this direction. Edgar Pieterse (2008, 28) describes post-1994 townships thus: "Old places, few jobs, youth cultures, soul of the so-called new South Africa, buzzy, vulnerable." These parameters are descriptive of the dire conditions that characterize the township politics (Mayekiso 1996). Yet millions of urban South Africans continue to live in these places. Pieterse's observation conveys the apostasy of the present-day township. It provokes us to reconsider what this apostasy means now, and, in the future, where referring to every urban scenario in post-1994 South Africa as "post-apartheid" may not provide an answer to the urbanisms underway. If the township of the past was interpreted as a signature of apartheid's racial bias, what does the contiguity of these conditions mean now? How do we name this present symbolism, which is so much like the past, yet officially detached from this past?

Jordache Abner Ellapen (2007, 116) thinks the "new" township is "a 'port' into major South African cities and a means of monitoring and controlling the urbanization of black South Africans." This observation alerts us to the shifting in the meaning of the township from apartheid paranoia to post-apartheid triviality. The township no longer signifies a physical location as such but is rather a euphemism for the condition of urban ephemerality associated with its uncertainties. Russell West-Pavlov (2014b) suggests that South Africa's post-1994 urban narratives are about urban ambivalence, proposing unintelligibility as a paradigm of post-apartheid urban South Africa. This paradigm is not far removed from the paradigm of the unknowable (Pieterse 2011), which prioritizes engaging with unforeseen urban parameters. Hillary Dannenberg's (2012) reading of Chris Abani's *Graceland* (2004) and Phaswane Mpe's *Welcome to Our Hillbrow* (2001) is exemplary of dynamic engagement with postcolonial African metropolises, in this case, Lagos and Johannesburg, respectively. Particularly, Hillary's analysis of Mpe's novel strengthens Russell West-Pavlov's (2014a, 12) reading of the same novel's layered urban space as indexical of a future underway. Per these readings of Africa's new urban imaginaries, self-identity has replaced racial identification (Frueh 2003), and theories of post-apartheid vagueness have usurped apartheid-era racial binaries as the fulcrum of South Africa's urban theory. To be vague, unknowable, unintelligible, or trivial is the new reality of the crisis of inhabiting the township. *Tsotsi* empowers this aesthetics of township.

An adaptation of Athol Fugard's novel by the same title, Garvin Hood's *Tsotsi* won the 2006 Best Foreign Film Academy Award. It tells the story of David (Benny Moshe), a township kid who flees his abusive and drunkard father (Israel Makoe) leaving behind a bed-ridden mother (Sindi Khambule). He grows up in another part of the township to become Tsotsi (Presley Chweneyagae), a gang leader. His gang is comprised of dim-witted Aap (Kenneth Nkosi), a killer named Butcher (Zenzo Ngqobe), and a school dropout named Boston (Mothusi Magano). *Tsotsi's* turning point is when, after a fight with Boston in the local shebeen (beer joints popular in South African townships), Tsotsi flees into the night to an upper-class black neighbourhood. Here, he robs a lone female driver, Pumla (Nambitha Mpumlwana), of her car outside the home she shares with her husband John (Rapulana Seiphemo). Shortly thereafter, as Tsotsi drives along the road, he discovers Pumla's baby (Nonthuthu/Nthuthuko Sibisi) inside the car. He takes the baby into the township and adopts it. The film's narrative is built around the family's effort to trace and recover their baby from the township, aided by security personnel.

But alongside this narrative, the film also curates an image of the black urban criminal colloquially named Tsotsi in the townships. That the film starts with a murder scene inside a public commuter train problematizes crime as an obsession of the township black youth (Kruger 2013, 65; Singh 2008). Thus, despite being made after apartheid, *Tsotsi* seems to inherit Fugard's characterization of the township as a crime zone. These two narratives converge in the character of Tsotsi, the protagonist who is portrayed as a criminal. The main crisis in the film is thus not physical (the effort to trace and recover the baby) or economic (the audacity to commit crime as a source of income); it is psychological (the incapacity of the township dweller to find a befitting identity). The main conflict concerns how to name the township by referring to a past of wanton cruelty (inherited from Fugard's novel) that is fully contrasted with the present of humaneness (the protagonist of the "new" township in this post-apartheid film is portrayed as empathetic and caring). Accordingly, if we see Tsotsi's character as an embodied urban form, then we can interpret his character as anthropomorphizing the post-1994 township's ambiguous future. He juxtaposes "an apocalyptic view and … an irrepressible optimism" (Pieterse 2008, 1) typifying the transitioning South African township as a space that is yet unknown. Apocalypse here infers a moment when apartheid has been replaced by socio-economic classes as the new segregation frontier (Gnad 2002), while optimism suggests the townships' struggle to invent a new symbol amidst rigid crisis.

This obscurity is evident in the conflicting reviews of the film. *Tsotsi* has been critiqued as a ghetto film genre shaping a "ghetto style and generic narrative formula that are detached from any political analysis of urban reality offered by urban studies and the lived experience of ghettos, barrios, and ethnic neighbourhoods" (Mennel 2008, 154). It has also been read as an aperture into the protagonist's "moral redemption" (Marx 2010, 266). Both interpretations recreate the township space as a "cinematic trope of blackness" (Ellapen 2007, 124), using the title of the film and the name of the protagonist as a euphemism for black gangsterism. These humanistic perspectives, while clearly emphasizing the protagonist's agency in assembling a conventional trope of post-apartheid township identity, disregard the textuality of nonentity conveyed through the mise-en-scène and cinematography of the "new" township, the main setting in *Tsotsi*. This chapter explores these two aspects as symbols of the emergent township identity. This argument is based on two aspects: first, that the urban mise-en-scène signifies an unintelligible township; and second, the township is a modular narrative space that expresses recursive marginalization. Both being aspects of *Tsotsi's* formal film visual style, they index nonentity as a paradigmatic label of the new South African township dweller.

Embodiment of nonentity in *Tsotsi*

After watching the hospital scene in which Pumla is recuperating, one may be intrigued by a rather casual dialogue between John, Captain Smit (Ian Roberts), Inspector Zuma (Percy Masemela), and a young police officer (Craig Palm):

JOHN: You said there were fingerprints?
CAPTAIN SMIT: This bastard's a nobody.
JOHN: But you said he went into the township?
INSPECTOR ZUMA: There's a million people in there. It's chaos.
YOUNG COP: Sir, we can't even track stolen cars.

That even with fingerprints – an official archive of biometric identity in any republic – and an illustrated image of Tsotsi drawn from Pumla's description, the police officers used epithets "bastard" and "nobody" to characterize Tsotsi usefully demonstrates the way *Tsotsi* deploys dialogue to augment its visual aesthetics of the township blacks as nobodies. Despite physically accessing the affluent city of middle-class blacks, Tsotsi remains outside the formal register of Johannesburg's residents both at the communal level (personified by John and Pumla's family) and the governance level (represented by the security officials.) Furthermore, his alienation is amplified through the contrast with John's family, which appears well-known to the police officers. Being unknown thus construes a metaphor of alienated urban existence that no longer relies on material inequality per se but rather originates from appropriation of obscurity as an isolating mechanism. In retrospect, this condition of existing as a nobody in Johannesburg anthropomorphizes the township, of which Tsotsi is emblematic, as a nonentity. It suggests a shift from the township as a space of abundant policing (Mbembe, Dlamini, and Khunou 2004, 502) or a space of "control, and hence of domination, of power" (Léfebvre 1991, 26), to a place "displaced even from itself" (Kupferman 2011, 142). Displacement here imbricates a visual eclipse of spaces whose structures would otherwise convey the close proximity of poverty to development and affluence, and by doing so, give agency to the discourse of urban segregation (Koonings and Kruijt 2009, 1). Through this erasure of racial, spatial, political, social, and economic symbols – otherwise noticeable in the squalid structures upon which its previous identity was sustained – and without acquiring new reference to express its post-apartheid frames, the township exemplifies the enigma typical of most post-apartheid symbols in South Africa. Such enigma is to be seen in the imaginaries of, for instance, the idea of a rainbow nation which somewhat suffers from the conjunction between racial cohesion and racial coexistence (Mututa 2019). In the township, after a near-complete reversal of apartheid's invasive surveillance, the resulting equivocality could thus be operationalized to designate an urban edge.

With its million residents collectively perceived as "nobody," the new township uses opaqueness to tactfully convey "utopias of effacement" (Boyer 2008, 55) as a key subtext in the film. The police officers' inability to identify

Tsotsi as a specific human being (and citizen) using his fingerprints, the epithet "nobody," and the suggestion that the township is his habitat outline the aestivation of post-1994 township perspectives already taking shape. While John's family represents the utopia of the black urban populace post-1994 and hence analogizes the "official" post-1994 Johannesburg of order, prosperity, and comfort, Tsotsi's gang metaphorizes the "reality of a vacant life" (Mbembe 2017, 6), the blossoming from an oppressive urban condition to a snubbed, unintelligible urbanite. This township appears as a vague space: neither typifying the racial order inscribed at the time of its establishment nor the new black identity typified by affluent black urban South Africans such as John's family. The cinematography of the shot occurring after Captain Smit, Inspector Zuma, and a young police officer discover Pumla's vandalized car strongly exemplifies how such nonentity is constructed within the film's visual aesthetics.

The shot is framed from a high angle and wide perspective. In the foreground, there is a tarmac road, with the vandalized car stationary on the pedestrian lane. These police officers stand beside the car looking towards the background, where we see an indistinguishable conglomeration of what appear as specks: the distant township. Except for the shimmer of dots, the township blends with the brownish scrub jutting from the dry withered landscape. The middle ground shows an open field with sparse shrubs. Noticeably, it is a scene of desolation. This use of perspective framing additionally obscures the details of the township infrastructure. Given that the township has historically been most discernible by its structural form of matchbox houses (Kruger 2013, 22), composing this perspective shot to render these structures unintelligible invites a more drastic reading of the township. It renders the township as a *"terrain vague* ... a space waiting to be cultivated with 'programmatic potential'" (Boyer 2008, 65). Privileging images of open barren landscape that emphasize its isolation visually "subtracts" the township from the proximity of the main city, personified by the police officers. It draws attention to the foreground-background axis as indexical of the spatial disconnection between the township dweller and the inhabitants of Johannesburg's formal city, who occupy the foreground. It is a visual aesthetic that invites us to notice, almost without effort, the world of the township as omitted from the usual territories with which the police officers on the foreground are familiar. Their hesitation to enter the township alludes to their responsiveness to this disconnect.

Further, the shot's use of a wide angle achieves vast horizontal exposure of the landscape, showing a subjective view of the township as distanced from the main city. This is coupled with occasional selective defocusing and blurring of the township in several medium close-ups, suggesting a visual quandary over how to represent the township through the perspective of a middle-class Johannesburg resident whose experiences are too far removed from the life of the township. The main issue raised in the shot's use of perspective is thus not merely spatial disenfranchisement but the attribution of unintelligibility to present a scenario in which the township appears to escape the post-apartheid urban grammar of inclusivity. It is in this terrain that the police officers insert the attributes of nonentity by branding the inhabitants as "nobodies," a tacit cipher for the inertness of the "urban unidentifiable."

Thus, a larger field of meaning is opened up by the term "nobody." Since it cannot be separated from the notion of post-apartheid Johannesburg as a split city, nobody speaks of the underlying organization of the city into the known (the connected and affluent) and the unknowns (the ignored and poor). For the township characters, being unknown is experienced as a sensation of "unreality ... a mere eye, a mocking intelligence, insubstantial and disembodied" (Wilner 1970, 256). Take Tsotsi, for instance, who overcomes the label of the brutal, beast-like protagonist by portraying a soft-hearted, humane personality. Yet, throughout the film, he is held in contempt and eventually punished. Such a plot, which subtly speculates that despite his transformation, he still "stays in touch with himself beneath the disguise he sells" (Ibid.), alerts us to the underlying anachronistic textuality of this construct of nonentity as a placeholder for post-apartheid urban binaries that no longer rely on racial articulation, but on being recognized as an urban resident or not. A probe into why Tsotsi's transformation is not acknowledged in the film is thus a probe of why his condition of unidentifiability is so crucial to the entire narrative of the township as a materially and socially unrecognized space. It is also an aperture into how lack of recognition operates to designate Johannesburg's post-apartheid "indifference to [Tsotsi's] eligibility" (Simone 2011, 1) in the new urban imaginary. Despite circulating between the township and the city, he remains a "nuisance ... a socioeconomic and sociocultural menace" (Gulema 2013, 185) to South Africa's urban expectations, post-1994. His subtraction from Johannesburg's modernity, implied by his personification of urban itinerants as the inhabitants of the township, actuates the crisis of the post-apartheid township in two ways: first, through the montage style and mise-en-scène of the township, which cast it as an unintelligible space; and second, through the noir and perspective cinematography, which tie the township's narrative to the modularity of historical sidelining.

Visualizing the unintelligible township

Tsotsi's first scene, a black screen overlaid with character voices, breaches a well-established cinema norm of beginning with an establishing shot to orient the viewer to the setting. These initial sounds, in the local dialect, are not transcribed with subtitles. To the viewer outside this linguistic community, this opening shot constitutes a moment where the film uses human language and visual style to conceal its message. A subsequent montage of interior medium shots further disorients the viewer to the spatial setting of the scene: a close-up of fingers resting on a worn tabletop, holding a smoking cigarette; a beer bottle next to the palm followed by interior shots of the protagonist's shack, where Aap and Butcher are playing dice, and Boston drinking beer and reading a newspaper; and a wide interior shot of all the characters, including Tsotsi, who is standing behind the other characters, his back to the camera. In the foreground, a bed is visible, the characters occupy the middle ground, and the background is a sunset sky with no discernible detail. The scene progresses with close-ups of dice, Butcher's rudimentary tool made of a long thin metallic shaft and a wooden handle, and shots emphasizing the characters' faces while downplaying that it is not divulging much about the setting. This visual style privileges the characters and mise-en-scène over the larger township, which, at this point, remains out of sight. In the absence of other details, the characters are metonymic of the absent township. Their "behaviors, attitudes and activities" are illustrative of "not only their own identity but also that of the (township) space" (Ellapen 2007, 120–121). If, then, we are to read the character of the township through mundane practices (Pieterse 2011, 11), such as drinking and gambling, these characters engaged in unproductive activities embody the new urban superfluous – bodies existing without goals, without duty, without meaning. They engender a thriving post-apartheid urban black character one might encounter in the streets of major South African cities today – blacks engaged in various provisional activities in strategic spaces such as road junctions, bus stops, petrol stations, or other such strategic locations – waiting for unknown jobs or economic windfalls for which they are neither qualified nor experienced. One also finds them at *robots* (traffic lights), often asking for financial assistance. These are the new faces of urban nobodies, the urban idlers who have been "forgotten" in the new social and economic dispensation in post-apartheid South Africa. Such are the faces intimated by the film's opening mise-en-scène. This group of strong and able youth idling without purpose anthropomorphizes a township no longer imagined through "exclusion, brutalization, and degradation" (Mbembe 2017, 6) but through being the "thing" that represents "nothingness" (Ibid., 2). In the post-apartheid context, this "nothing" prefigures a lack of proper urban status or agenda. The township constructed here replaces the physical space with atypical urban practices.

But if we are to grasp the intricacies of the nature of crisis implicit in such a turn of events, and which the everyday urban life of the township dweller expresses, we must become attentive to the way the film language is working at a semiotic level. Here, everyday experiences of lack suggest absence of sufficient support mechanisms and opportunities for the township dweller. It is on this basis that we can read the mise-en-scène and composition of the first exterior shot of Tsotsi's gang leaving his shack as quintessential of the township. Here, the characters occupy the foreground surrounded by a vista of rusty rooftops comprising indistinguishable low-lying structures to the right and left lower half edges of the shot's middle ground. The background, framed by the upper vertical half of the screen, shows a late evening sky blending with the shot's sepia tone. Visually, the shack is framed to appear larger in proportion to the other elements in the frame. It is positioned at the foreground centre, elevated vertically above the rest of the indistinguishable township (the maze of rooftops in the middle ground), visually isolating it as the shot's primary symbol, rendering the characters as secondary symbols. This framing, which omits the township's landscape of vast conglomeration of squalid structures by use of a low camera angle while giving prominence to the characters, creates a dramatic instance in which we contemplate the contemporary township through the character of its inhabitants rather than via material symbols. That is, the township's material symbolism is replaced by the "character" of its residents. General attributes of the key supporting characters include illiteracy and semi-literacy, habitual criminality, excessive poverty, ignorance, and drunkenness. These attributes constitute barriers that keep these characters outside the main city.

However, through Tsotsi's character, the film positions this barrier as significant in assigning unintelligibility to the post-apartheid township. If we probe beyond the paradox occasioned by his assigned identity of criminality, and his manifested identity of humaneness, we arrive at a contradiction between the past and the future. The former reifies a stereotypical reproduction of the township while the latter reifies a resistance to this stereotype. At a hermeneutic level, then, he caricatures a historical turning point, where the past urban identity has been turned "inside out" (West-Pavlov 2014a, 9) – meaning he now stands as a character with no clear identity. Consequently, subsequent images of Tsotsi's shack – the site of his inner transformation from exemplifying urban crime culture to humane struggles for survival – challenge the historical imaginary of the township it mimics materially. It denotes that which is "continually unnamed, that which can be neither symbolized nor mastered" (Benjamin 2010, 27). It compels us to imagine this shack as challenging the customary epithets of Tsotsi urban culture and to notice its embeddedness within the township as a visual grammar of a space and its occupants being re-imagined and reinterpreted as markers of defunct urban vocabulary. If Tsotsi's daily practices suggest a certain type of post-1994 township character abandoned by the incumbent development policies, his moral vagueness (Pieterse 2011, 12), here actuated by his characterization as neither a criminal nor an acceptable urban dweller, alerts us to a less-acknowledged scenario: that the post-1994 township has outgrown its historical label without acquiring an alternative one. It can be described as a "border zone where the subjective and the material collide" (Ibid., 18), that is, a nonentity.

The township as a modular space

Tsotsi's first orientation shot of the township, complete with its vast structural maze, is set at night, and dominated with low-key shadowy visuals. The camera is positioned high, shifting emphasis from the minute characters to the darkly lit shacks. This shift in camera position from low to high, and in mise-en-scène from characters to the overall township structures, accompanied by a shift in lighting from the sepia tone of the sunset seen earlier to the dark hues of the night, produces an aesthetic of shadows. These shadows may be read in Ruth Kerkham Simbao's (2007, 58) reasoning; as allowing for "uneven ground where both dominant and counter-narratives might exist." Per this interpretation, *Tsotsi's* use of film noir, an established film genre that emerged after World War II cinema and uses low-key visual lighting style among other visual aesthetics, is a possible aperture for espying the ambiguities of the post-1994 township imaginary in cinema.

This subsection argues that *Tsotsi's* use of noir visual style to depict the township space simultaneously conceals the township's dereliction. In effect, it renders the topography nondescript, enticing the viewer to contemplate the meaning of such visual style. One way to read this noir visual style is that while it identifies minuscule details – such as the setting and the street lamps – thus mapping the township's "spatial memory" of apartheid surveillance (Graham 2009, 1–2), it nevertheless obscures the township beyond the immediate topography of box-like structures visible in the chiaroscuro lighting. What the viewer sees is a nondescript space, a spatial nowhere. Despite the wide-angle framing of this establishing shot, the noir style generates ambiguity about whether the shot helps to "reveal or conceal ... important information or background to the audience" (Brown 2012, 10). It poses the township as an inscrutable symbol that requires further investigation to comprehend its meaning. This reading is further supported by the fact that the shot lacks any focal element, instead appearing as a happenstance of the township rather than a framing of a specific township space. Accordingly, the metaphor of an unfurnished space indexed by this framing and the noir lighting instantiates a moment when the film seems to lack an appropriate visual concept of the post-1994 township and hence deploys, from within its visual toolset, noir lighting and nondescript mise-en-scène as opaque signifiers. Extending the idea of Tsotsi as an embodiment of the township, the scene intimates a moment when the post-1994 township of *Tsotsi* becomes inexpressible. The critical question precipitated by such cinematography is: what dramatic nuances of the township does this noir visual style render, and what do they intimate about the new township? This may be answered through the parallel subplot of Tsotsi's personal life, which historicizes the township with a flashback of his childhood. This narrative style creates a parallel between the township as an experience (of disease, poverty, cruelty, and inhumane difficulties) and the township as a space (of neglect, hustle, escapism, and defeat). The visual gaps occasioned by the indistinguishability of the noir township, when read alongside Tsotsi's "unknown" personal life, "foreground the relationship between the temporality of the story and the order of its telling" (Cameron 2008, 1). I term this aesthetic a "modular narration."

Temporality is concerned with whether a film tells a linear narrative following a natural rhythm of time, while the order of narration is concerned with montage choices and juxtapositions of events and places. In the case of *Tsotsi*, temporality is a lens through which to consider if the film recognizes the new township as a valid new symbol detached from the accoutrements of its history or only reiterates this historical symbolism. It also concerns, first, the past–present interconnection of the story, which draws on the adaptation of a 1960s novel into a 2005 film; and second, the cyclic experience of neglect in the township.

On temporality, the gangster genre links Fugard's 1960s township narrative with Hood's 2005 translation of the same. Through the lone marginal figure of Tsotsi and his presentation as a township gangster, we feel a sense of repetitive history. The word used as the title, "Tsotsi," which translates to urban youth criminal, popularly used in the 1960s, draws a connection between the historical trope of gangsterism and appropriates that trope to express a 2005 township. While the genre links the two versions of the township, it stirs a sense of inarticulacy around the psychological changes evident in the protagonist's life.

Tsotsi opens with a murder scene and moves on to a shebeen fight scene, characterizing Tsotsi as ruthless. It ends with a scene where Tsotsi – who took great risks to care for the baby he found and who even kills Butcher, a gang member, to save the life of John, a humane character who fought against crime – is being arrested. Disallowing his goodness and obscuring it with the gangster identity as the film does in the final scene privileges a stereotypical identity of criminality. Having seen Tsotsi's effort to save the baby (these remain unknown to the security officers and the parents), the film adduces that Tsotsi is not judged on his present character, but on his presumed past. Through him, the new humane township that he embodies is usurped by the township of the past. The confrontational ending scene relays this problematic scenario where summoning the past imaginations of township youth potentially replaces the psychological gains attained by the protagonist in projecting a humane character, thus limiting his capacity to override these stereotypes. It re-enacts a commonplace ritual of fascination with "seeing" the post-1994 township dweller as an empty signifier – hence the use of the terms "bastard" and "nobody" to refer to Tsotsi. All this happens alongside the enigma of how to represent the new township without affecting the structure of a filmic narrative that privileges a static history.

Starting with the image of Tsotsi as a gang leader while incorporating flashbacks of his traumatic childhood, then remodelling his childhood through John's baby (whom he names David, after his childhood name), and finally ending with the adult Tsotsi under arrest for delivering the baby safely back to its parents – the film's structure uses the temporality of the character's narrative arc and his psychological arc to invoke a Hollywood modular narrative structure in which the ending is already given (Cameron 2008, 55). He appears predetermined as a township thug, whether through accentuating bad actions that support this recognition or concealing good actions contradicting this perception. Such a narrative invests in specific materiality (Mbembe 2001, 5), namely, validation of the protagonist's insignificance. The film fails to conclude whether Tsotsi is really the thing that represents the imagined township of gangsters, its double – the post-apartheid obscurity of urban blacks or just a sign that is not a sign at all (Ibid., 142). By centring this film on such a protagonist, the film relies on stagnated historical temporality and cinematographic conventions to convey the obscurity of the new township within Johannesburg's post-apartheid signifiers. That the film is adapted from a novel that centres on accentuating criminality may explain the narrative's positive inclination towards rendering a static perception of the township as a space that did not quite outgrow its historical label.

On the order of narration, the inability to determine accurately the attributes of the township being inferred through characterization of its residents and visualization of the setting are the main aspects through which we gain knowledge of the township being imagined and constructed within *Tsotsi*. A cursory dialogue that precedes Tsotsi and Boston's scuffle inside the shebeen is a useful starting point:

BOSTON: What's your name, Tsotsi? Your real name? Tsotsi? Thug? That's
    not a real name.
AAP: Boston, don't ask him questions.
BOSTON: Why not? I've been with you for six months. I want to know his
    name. Every man has a name. A real name from his mother.

At the outset, this altercation signals Tsotsi's opaqueness, even among his colleagues. It nevertheless also represents an attempt to crack open the protagonist's linear past–present story (which his colleagues do not know) and the non-linear narrative structure used in the making of *Tsotsi* (with Tsotsi's flashbacks and Morris' backstory contributing significantly to the meaning of the film). When Boston says: "I want to know his name. Every man has a name. A real name from his mother," he is pre-empting a retrieval of Tsotsi's mysterious history. It is noteworthy that this scene is located within the opening sequence – in diegesis time, occurring within a few minutes of the film story's start, and thus suggesting an urgency to express the protagonist's opaqueness.

Thus, the subsequent scene juxtaposing shots of Tsotsi fleeing through veld in total darkness, with shots of a similar moment in his childhood as David (the young Tsotsi) fleeing his home – collocates the two moments beyond chance: they are seen as constitutive of a repetitive experience of the township. The sporadic bursts of lightning, the rain, the noir light, and dark hues create a temporal link between these two moments. When the adult Tsotsi halts under a tree in a middle-class neighbourhood, the film cuts to the young Tsotsi inside a concrete culvert, the rain pounding heavily around him. It is the first time the viewer gets to know Tsotsi's present as a mirror of his past. The concurrence of Tsotsi's past and present in this montage is an apposition that indicates a "present frozen in a kind of stasis … a time and place where what happened is still happening, just as the present is happening now" (Hoy 2009, 95). The montage elicits a recursive narrative that links the present township of Tsotsi's youth to the past township of his childhood – not so much as to distinguish between the two, but to express the unknowability of the present township. His persistent misery across this historical period – from his childhood family characterized by drunkenness, sickness, and poverty, to his adulthood gang community characterized by illiteracy, poverty, crime, drunkenness, and violence – can be understood in later sequences of the film when Tsotsi visits the open field where he once lived, now occupied by younger generations of township kids. Not only do these kids mirror his past, they also pre-empt the question: what has the township become in the present times?

Boston's question, "What's your name, Tsotsi?," thus positions the film as an anachronic narrative that not only intersperses the simultaneity between the apartheid past and the present but also construes the township as a "temporally fractured but ontologically unified world" (Cameron 2008, 141). It thus alerts us to a "protagonist who is not so clearly defined as purely good or evil" and "characters full of contradiction and alienation" (Brown 2012, 69), and synchronizes his status as nobody, initially symbolized by existence in the open veld, with his present unintelligibility in the city's formal register. In this perspective, Tsotsi the character and Tsotsi the embodied township stand in for a concern with post-apartheid urban displacement and the uncertainty of what the township can or has become.

This preoccupation with the uncertain future of the township may be read through the character of Morris (Jerry Mofokeng), the crippled beggar. We first encounter Morris at Johannesburg Park Station. Tsotsi's encounter with Morris happens after Tsotsi accidentally steps on Morris, who in turn insults him – calling him a dog. The rest of the sequence is built around Tsotsi pursuing Morris as he pushes himself in a wheelchair across the city while other members of Tsotsi's gang await him at the train station where they had planned to commit a crime. The most remarkable visual attribute of this sequence is the setting and the lighting: in the lonely space beneath a bridge, Tsotsi subdues and coerces a backstory from Morris, who explains that he was crippled in an accident while working in a goldmine. It is a moment where the distant past (signified by the age gap between the characters) being replayed augments the subtext of recursive history as very much the present and the future.

The other time we encounter Morris is also at Johannesburg Park Station, in the film's final sequence, when Tsotsi is on his way to returning John's baby. Here, Tsotsi gives Morris money and proceeds on his journey to John and Pumla's home. In this moment, Morris's static lifestyle of begging embodies economic fixation, while Tsotsi personifies criminality – the attribution fixed to the township and its residents. In this respect, both characters embody a township "where the horizons of a reasonably attainable future and the capacity to imagine them have disappeared" (Simone 2004, 3). *Tsotsi*, it would seem, is not invested in a post-1994 township that is progressing towards inclusion but keeps on recycling past tropes (at this point of economic stagnation and a stereotypical labelling of criminality) to signal the increasing worthlessness of the township as a fixed, inescapable crisis. Going forward, the questions would be: in what ways, if any, does the reliance of John's world on the obliteration of Tsotsi's world to exist constitute not just a combative process of township visual imagination, but premiers a narrative style that potentially posits the post-1994 township as a figurative sign, a textuality of nonentity? That is, how can we conceptualize nonentity as a crisis, and what value would such a concept offer in theorizing urbanism in post-apartheid Johannesburg?

Bibliography

Abani, Chris. 2004. *GraceLand*. New York: Picador.
Badcock, Blair. 2002. *Making Sense of Cities: A Geographical Survey*. London: Arnold.
Baines, Gary. 2007. "The Master Narrative of South Africa's Liberation Struggle: Remembering and Forgetting June 16, 1976." *The International Journal of African Historical Studies* 40 (2): 283–302.

Benjamin, Andrew. 2010. "Trauma Within the Walls: Notes Towards a Philosophy of the City." *Architectural Design* 80 (5) (Post-Traumatic Urbanism September/October): 24–31.

Boyer, Christine M. 2008. "The Many Mirrors of Foucault and Their Architectural Reflections." In *Heterotopia and the City: Public Space in a Postcivil Society*, edited by Michiel Dehaene and Lieven De Cauter, 53–74. London and New York: Routledge.

Brown, Blain. 2012. *Cinematography - Theory and Practice: Imagemaking for Cinematographers & Directors (Second Edition)*. Waltham and Oxford: Elsevier.

Brown, Julian. 2016. *The Road to Soweto: Resistance and the Uprising of 16 June 1976*. New York: James Currey.

Cameron, Allan. 2008. *Modular Narratives in Contemporary Cinema*. Hampshire and New York: Palgrave Macmillan.

Canepari, Michela. 2011. "Visible and Invisible Borders in Post-Apartheid South Africa: an Intersemiotic Analysis of Meta-History." In *Oltre i confini. Testi e autori dell'esilio, della diaspora, dell'emigrazione*. Vol. 1, 217–260. Parma PR: Monte Università Parma.

Dannenberg, Hilary. 2012. "Narrating the Postcolonial Metropolis in Anglophone African Fiction: Chris Abani's GraceLand and Phaswane Mpe's Welcome to Our Hillbrow." *Journal of Postcolonial Writing* 48 (1) (February): 39–50.

Dovey, Lindiwe. 2009. *African Film Adapting Violence to the Screen*. New York: Columbia University Press.

Duminy, James, Susan Parnell, and Mercy Brown-Luthango. 2020. *Supporting City Futures: The Cities Support Programme and the Urban Challenge in South Africa*. Cape Town: African Centre for Cities.

Ellapen, Jordache Abner. 2007. "The Cinematic Township: Cinematic Representations of the 'Township Space' and Who Can Claim the Rights to Representation in Post-Apartheid South African Cinema." *Journal of African Cultural Studies* 19 (1) (June): 113–137.

Frassinelli, Pier Paolo. 2015. "Heading South: Theory, Viva Riva! and District 9." *Critical Arts* 29 (3): 293–309.

Frueh, Jamie. 2003. *Political Identity and Social Change: The Remaking of the South African Social Order*. Albany: State University of New York Press.

Gnad, Martin. 2002. *Desegregation und neue Segregation in Johannesburg nach dem Ende der Apartheid*. Kiel: University of Kiel.

Graham, Shane. 2009. *South African Literature After the Truth Commission: Mapping Loss*. New York: Palgrave Macmillan.

Guelke, Adrian. 2005. *Rethinking the Rise and Fall of Apartheid: South Africa and World Politics*. Hampshire and New York: Palgrave Macmillan.

Gulema, Shimelis Bonsa. 2013. "City as Nation: Imagining and Practicing Addis Ababa as a Modern and National Space." *Northeast African Studies* 13 (1): 167–213.

Hart, Deborah. 1986. "Literary Geography of Soweto." *GeoJournal* 12 (2) *South Africa: Geography in a State of Emergency* (March 1986): 191–195.

Hirson, Baruch. 1979. *Year of Fire, Year of Ash: The Soweto Revolt: Roots of a Revolution?* London: Zed Press.

Hoy, David Couzens. 2009. *The Time of Our Lives: A Critical History of Temporality*. Cambridge, MA and London, England: The MIT Press.

Koonings, Kees, and Dirk Kruijt. 2009. *Megacities: The Politics of Urban Exclusion and Violence in the Global South*. London and New York: Zed Books.

Kruger, Loren. 2013. *Imagining the Edgy City: Writing, Performing, and Building Johannesburg*. New York: Oxford University Press.

Kupferman, David W. 2011. "On Location at a Nonentity: Reading Hollywood's 'Micronesia'." *The Contemporary Pacific* 23 (1): 141–168.

Léfebvre, Henri. 1991. *The Production of Space*. Translated by Donald Nicholson-Smith. Oxford: Blackwell.

Léfebvre, Henri. 1996. *Writings On Cities*. Edited by Eleonore Kofman and Elizabeth Lebas. Translated by Eleonore Kofman and Elizabeth Lebas. Oxford: Blackwell Publishers.

Marx, Lesley. 2010. "At the End of the Rainbow: Jerusalema and the South African Gangster Film." *Safundi: The Journal of South African and American Studies* 11 (3): 261–278.

Mayekiso, Mzwanele. 1996. *Township Politics: Civic Struggles for a New South Africa*. New York: Monthly Review Press.

Mbembe, Achille. 2001. *On the Postcolony*. Berkeley/Los Angeles/London: University of California Press.

Mbembe, Achille. 2017. *Critique of Black Reason*. Translated by Laurent Dubois. Durham and London: Duke University Press.

Mbembe, Achille, Nsizwa Dlamini, and Grace Khunou. 2004. "Soweto Now." *Public Culture* 16 (3): 499–506.

Mennel, Barbara. 2008. *Cities and Cinema*. London and New York: Routledge.

Mpe, Phaswane. 2001. *Welcome to Our Hillbrow*. Pietermaritzburg: University of Natal Press.

Mututa, Addamms Songe. 2019. "Johannesburg in Transition: Representing Street Encounters as Racial Registers in Clint Eastwood's Invictus." *Critical Arts South-North Cultural and Media Studies* 33(2): 1–13.

Mututa, Addamms Songe. 2020. "Customizing Post-Apartheid Johannesburg: The Dialectic of Errancy in Ralph Ziman's Jerusalema." *Safundi: The Journal of South African and American Studies* 21 (2): 206–223.

Nuttall, Sarah. 2006. "A Politics of the Emergent Cultural Studies in South Africa." *Theory, Culture & Society* 23 (7–8): 263–278.

Pieterse, Edgar. 2008. *City Futures: Confronting the Crisis of Urban Development*. Lansdowne: UCT Press.

Pieterse, Edgar. 2011. "Grasping the Unknowable: Coming to Grips with African Urbanisms." *Social Dynamics: A Journal of African Studies* 37 (1): 5–23.

Rogerson, Christian M., and Maria Da Silva. 1988. "From Backyard Manufacture to Factory Flat: the Industrialisation of South Africa's Black townships." *Geography* 73 (3) (June): 255–258.

Simbao, Ruth Kerkham. 2007. "The Thirtieth Anniversary of the Soweto Uprisings: Reading the Shadow in Sam Nzima'sIconic Photograph of Hector Pieterson." *African Arts* 40 (2) (Summer): 52–69.

Simone, AbdouMaliq. 2004. *For the City Yet to Come: Changing African Life in Four Cities*. Durham and London: Duke University Press.

Simone, AbdouMaliq. 2011. "The Ineligible Majority: Urbanizing the Postcolony in Africa and Southeast Asia." *Geoforum Geoforum* 42: 266–270.

Singh, Anne-Marie. 2008. *Policing and Crime Control in Post-apartheid South Africa*. Hampshire: Ashgate Publishing.

West-Pavlov, Russell. 2014a. "Inside Out – The New Literary Geographies of the Post-Apartheid City in Mpe's and Vladislavić's Johannesburg Writing." *Journal of Southern African Studies* 40 (1): 7–19.

West-Pavlov, Russell. 2014b. "Shadows of the Past, Visions of the Future in African Literatures and Cultures." *Tydskrif Vir Letterkunde* 51 (2): 5–17.

Wilner, Eleanor R. 1970. "The Invisible Black Thread: Identity and Nonentity in 'Invisible Man'." *CLA Journal* 13 (3) (Special Ralph Ellison Number/March): 242–257.

Filmography

Battle for Johannesburg. Dir. Rehad Desai, South Africa, 2010.

Cape of Good Hope. Dir. Mark Bamford, South Africa, 2004.

Chikin Biznis - The Whole Story. Dir. Ntshaveni Wa Luruli, South Africa, 1999.

Conversations on a Sunday Afternoon. Dir. Khalo Matabane, South Africa, 2005.

District 9. Dir. Neill Blomkamp, USA, 2009.

Gazlam. Dir. Alex Yazbek, South Africa, 2002.

God Is African. Dir. Akin Omotoso, South Africa, 2001.

Hijack Stories. Dir. Oliver Schmitz, South Africa, 2000.

Invictus. Dir. Clint Eastwood, South Africa, 2009.

Jerusalema, Gangsters' Paradise. Dir. Ralph Ziman, South Africa, 2008.

Jump the Gun. Dir. Les Blair, South Africa, 1997.

Max and Mona. Dir. Teddy Mattera, South Africa, 2004.

Miners Shot Down. Dir. Rehad Desai, South Africa, 2014.

Proteus. Dir's. John Greyson and Jack Lewis, South Africa, 2003.

Rape for Who I Am. Dir. Lovinsa Kavuma, South Africa, 2005.

Tengers. Dir. Michael J. Rix, South Africa, 2007.

Tsha Tsha. Dir's. Shaft Moropane and Mosese Semenya, South Africa, 2003.

Tsotsi. Dir. Garvin Hood, South Africa, 2005.

Wooden Camera. Dir. Ntshaveni Wa Luruli, South Africa, 2003.

Yizo Yizo. Dir's. Barry Berkc, Andrew Dosunmu, Angus Gibson, and Teboho Mahlatsi, South Africa, 2004.

Zulu Love Letter. Dir. Ramadan Suleman, South Africa, 2004.

3 *Laissez faire* urbanism
Economies of dystopia in postcolonial Kinshasa

Postcolonial Kinshasa

Like many postcolonial African cities, Kinshasa has evolved towards informality. This is most evident in the city's proliferation of informal street economies. In the city's periphery, these feed into the perception of popular practices of informal survival in postcolonial urban Africa as not "constrained by its material conditions" but instead offering "something no legal neighborhood could: freedom" (Neuwirth 2005, 5). Street vending anchored in individual enterprise offers a template to speculate on futile urban survival, and hence acts as a provocateur of the shadowy forms of the urban that spring up precisely in hotspots of urban crises. This chapter discusses the economies of dystopia in postcolonial Kinshasa. Labelling the resulting shadowy city life – a space where established procedures guiding urban life in many formal spaces are hardly applicable (Simone 2007a, 84) – as "*laissez faire* urbanism," this chapter underscores the increasing centrality of unconventional economic practices in characterizing Kinshasa's postcolonial urban trends. Through a close reading of Djo Tunda Wa Munga's representation of Kinshasa in *Viva Riva!* (2010), this chapter illustrates the way individual-based survivalist entrepreneurship, labelled "*laissez faire* economic practices*,*" becomes a placeholder for Kinois' guarded mode of urban survival. The chapter taps into existing local terminologies circulating in Kinshasa – *Khadaffi* (the entrepreneurs who hoard fuel to sell it at high profits) and *la sapeur* (Kinshasa residents who dress up in expensive clothes to give impression of opulence) – to illustrate how the perilous economic practices in the face of ever-present uncertainty, which they index, engender circumspection as an ingenious response to urban economic crises.

De Boeck and Plissart (2004, 79) pose an important question: "Is all that remains in the postcolony, the post mortem, an autopsy of crisis?" This attribution of post-death procedures to postcolonial Africa presumes an unsalvageable breakdown of political, social, and economic systems necessary for a nation's existence. In the postcolonial Democratic Republic of the Congo (DRC), an autopsy could be closely linked to the contiguous narratives of "backwardness, barbarism, incompetence … and savagery" (Dunn 2003, 109) that permeate the country's history of mineral extraction, wars, disease, and famine. In Kinshasa, the term "autopsy of crisis" evokes an exhausting urban

DOI: 10.4324/9781003122098-3

life, devoid of the essential provisions necessary for survival or recognition within the formal urban structures that sustain urban life. Incessant conflicts and wars have resulted in a perpetual scenario of doing without. Throughout Kinshasa's postcolonial history, citizens have had to do without "food; they do without fuelwood; they do without primary health services; they do without safe drinking water. They also do without political participation, security, leisure or the ability to organize their time as they would like" (Trefon 2004, 4). In the popular dialect Lingala, this situation is designated *tokokufa*, literally translating to "we are dying" (Ibid., 4). While this trope draws from a national-scale economic condition, it easily maps out the desperation with which residents of Kinshasa approach various crises that comprise the city's daily dystopia. The question at hand is not about the state of affairs but about adaptive behaviours that become necessary to exist within such conditions.

This calls attention to the peculiar scenario in Kinshasa's neighbourhoods, where those continually struggling to exist outside legal or procedural legislation, and where "incessant checking and conversations, become the mechanisms for keeping things from getting out of hand" (Simone 2011, 333). In the absence of formal interventions and regulations, Kinshasa's edge has "bypassed, redefined, or smashed the colonial logics that were stamped onto its surface ... spatially, in terms of its architectural and urban development, as well as in terms of its socio-cultural and economic imprint" (De Boeck 2011, 313). For a city so embroiled in religious Pentecostalism (Pype 2012), such "smashing" practices are veiled under the everyday economic enterprises embodied by the "Khadaffi" and "la sapeur," both popular Kinois' identity practices. This chapter critiques the representation of these characters in Djo Tunda Wa Munga's film, *Viva Riva!* (2010) as rendering a dystopic, individual-based view of *laissez faire* economic models.

Viva Riva! is narrated around a fuel crisis in Kinshasa. The main character, Riva (Patsha Bay Mukuna), arrives in Kinshasa from Angola with boatloads of fuel during a severe fuel shortage. Riva's business associate G. O. (Romain Ndomba) inhabits the city's edge alongside a friend, J. M. (Alex Hérabo), and a street boy, Anto (Jordan N'Tunga). Meanwhile, a violent Angolan criminal gang led by Cesar (Hoji Fortuna) arrives in the city wanting to repossess the fuel – the main conflict driving the film. Working alongside a former prostitute, Malou (Angélique Mbumb), and military official, a La Commandante (Màrlene Longange), the gang commits murders across Kinshasa while searching for Riva. Meanwhile, Riva falls in love with Nora (Manie Malone), the girlfriend of a celebrated sapeur, Azor (Diplôme Amekindra), introducing a secondary conflict about contested masculinity (Pype 2007) in the film. Juxtaposing two versions of identities – Azor, la sapeur, a refined urban character known for his stylish dress and mannerisms, and Riva, *la lutteur*, a struggling urban character known for aggressive behaviour (Bazenguissa-Ganga 2014, 171) – the film weaves an intricate plot of violence, murder, poverty, pleasure, and betrayals. The argument pursued here is that the entwinement of these two identities, both of which explicate peculiar responses to Kinshasa's economic difficulties, characterize Kinshasa's dystopic postcolonial urbanism.

At the seventh Africa Movie Academy Awards ceremony held at the Glory Land Cultural Centre, Nigeria, on 27 March 2011, *Viva Riva!* won the 2011 MTV Movie Awards for Best African Movie and Best Picture, Best Director, Best Actor in Supporting Role, Best Cinematography, and Best Production Design. As such, *Viva Riva!* commands much attention in contemporary narratives of Kinshasa, drawing the viewer into survivalist economic practices as current concerns in postcolonial Kinshasa. Through this verisimilitude of urban distress among Kinshasa's informal urban residents, Wa Munga claims a place among local and global film audiences interested in urban citizenship in postcolonial Kinshasa. He joins other postcolonial filmmakers in Kinshasa, including Mwezé Ngangura, Balufu Bakupa-Kanyinda, and Raoul Peck. Other notable participants in DRC's postcolonial filmmaking include international filmmakers such as Roland Joffé's *City of Joy* (1992), which premiered on Netflix; Daniel McCabe's *This is Congo* (2017); and *Congo in Four Acts* (2010), co-directed by Dieudo Hamadi Kiripi Katembo and Divita Wa Lusala. Dieudo Hamadi is also well-known for *Kinshasa Makambo* (2018), a political activist film set in Kinshasa.

Some reviewers critique *Viva Riva!* as exemplifying the intersection of digital media platforms and new viewing habits in new popular African cinema (Tchouaffe 2016). Chris Chang (2011) talks of the film as a gangster genre film propelled by "ticking-time-bomb tension that propels so many American genre flicks," acknowledging the centrality of the fuel business in the same review: "the liquid is so precious that the siphoner spits his mouthful into the container." Pier Paolo Frassinelli (2015, 299) describes the film as showing Kinshasa "detached from the past and represented as its own social and political failure." Contrarily, this chapter draws its arguments from Karen Bouwer's (2019, 70) reading of the film, which focuses on a "continuum of mobility: from inertia (forces of stagnation and death) to flow (not simply motion but also moments of possibility and affirmation of life)." In consideration of its key concerns with economic structures and the use of personal urban experiences as the main narrative material, this chapter critiques the informal financial model built around the illegal fuel business as indexical of the way practices of individual-based economic practices in Kinshasa continue to impact the city's survival infrastructures of flow in the present times. The chapter focuses on the representations of two existing, local street-smart identities circulating in Kinshasa, Khadaffi and la sapeur. Although these identities easily fit Chang's (2011) description of the representation of Kinshasa streets as "suffused with gritty angst and palpable desperation," they most conveniently elicit ideas of opportunism such as he envisions in his description of the film's protagonist as a "man who thrives on scarcity." This chapter picks up from this cue and argues that both the Khadaffi and la sapeur, dominant identities in this film, illustrate the inventiveness of *laissez faire* economic responses to postcolonial dystopia in Kinshasa.

Laissez faire **urbanism**

Like many postcolonial African cities where the majority of residents live in informal conditions (Pieterse 2011), Kinshasa has evolved towards informality. Here, "illegally developed plots outnumber the legal ones by ten-fold" (Simone 2004, 196), which accounts for the city's slum-like sprawl. The practice of illegality and informality is also pervasive in DRC's street economies (De Herdt and Marivoet 2018, 121), where inventiveness has become a normal way of life. For instance, Joe Trapido (2016) notes the commercialization of music, whereby a love song dedication may cost thousands of dollars; while Bob White (2008) explores the adoption of *libanga*, the pronouncement of a wealthy person's name during performances in exchange for financial support. Such entrepreneurship pervades almost all aspects of Kinshasa's informal economies. For street entrepreneurs such as

> *romains* (dealers in smuggled merchandise), *bana kwatas* (touts dealing in second-hand clothes), *chayeurs* (wholesalers' agents), *gaddafis* (informal fuel-sellers), *chargeurs* (touts working for taxis and public transport, not to be confused with battery chargers!), *cambistes* (street money-changers), and *mamas manoeuvre* (middlemen trading food products in river ports)"
> (Ayimpam 2019, 13)

informality is often the only available chance for survival. The proliferation of informal street economies and the economic difficulties they address in people's daily subsistence means that those reliant on such economies have to adopt non-conventional patterns of inhabiting the city. Such peculiar patterns of postcolonial urbanism are based on individual profiteering, negating the possibility of the largely absent government to participate or intervene. For most of Kinshasa's street economies – and the edge spaces inhabited by the majority of informal traders – the government has let things take their own course.

Consequently, a self-regulating economic arrangement based on "hoarding and speculation" (Simone 2004, 196) has taken root. While this is of course not new, usefully engaging with such a scenario interpolates Kinshasa's everyday with the city's mode of urbanism. If Kinshasa's informality and squalor indicate insufficient government involvement in the provision of necessary survival infrastructure (Pieterse 2011, 8), then individual-based survival is a response to prevailing socio-economic crises. Free enterprise and dire survival are not mere indicators of an absence of government involvement but signify an in situ informal policy. Arguably, instead of appearing as unwillingness to fully meet its obligations to its citizens, the government has "authorized" urban residents – through its absence – to pursue individual livelihoods as they see fit. From this tacit arrangement, a curious form of economic urbanism takes shape. Out of surviving in this off-grid life, Kinshasa's edge citizens have become "actors able to buy themselves out of the cityness of city life – out of its thickening, unpredictable intersections of things of all kinds" (Simone 2007b, 237). Terming the personalized urban experiences of Kinshasa's economic devastation as *laissez faire* urbanism, this chapter argues that the resulting dystopia has curated peculiar individualities in Kinshasa. It is a close reading of *Viva Riva!*'s representation of dominant modes of urban dystopia as a response to *laissez faire* economic practices in postcolonial Kinshasa.

Adam Smith, a political economist and philosopher from Scotland, theorized *laissez faire* economic practice as a free-market model. He advocated for self-regulated economic practices, suggesting that governments should avoid market regulation (Smith 1776). In postcolonial Africa, this model has promoted self-centred interests leading to dreadful urban conditions. Dystopia suitably captures Kinshasa's chaotic economic condition, which has since independence been tied to anarchic politics. *Viva Riva!*'s narrative can be termed as a criticism of the *laissez faire* economic model. The characters' economic hardships suggest the failure or shortcomings of individual profiteering amid a lack of self-regulation. For all the characters, most of their economic activities provide neither profit nor lift them from persistent destitution (Iyenda 2005, 67). How do we interpret a city where livelihoods are not guaranteed but rather are achieved through speculation? This question alerts us of the paradox of the imagined and the lived Kinshasa in the film, which is concurrently a paradox between the imagined success of the *laissez faire* economic model and its catastrophic consequences. This is exemplified by the conflict between Riva's imaginary of opulence, trust, happiness, and social vibrancy and the verisimilitude of violence, crime, murder, poverty, and general malaise.

Viva Riva!'s opening sequence is quite remarkable in characterizing the dystopic outcomes of *laissez faire* urbanism in Kinshasa. In the opening shot, we see throngs of people walking along a dilapidated street located centrally in the screen. There are informal businesses on either side of the street, marked by a cascade of umbrellas. As the scene advances, we also see some cart pushers within the street crowds. From the composition of the shot, the street is outside the main city of Kinshasa, La Ville, which is just visible in the background. The prominence given to the walking crowds and the informal businesses mimics the arduous rhythm of daily existence. Lack of proper public mobility infrastructure, intuited by the walking masses, echos De Boeck and Plissart's (2004, 94) metaphor of "line II" or "foot bus," a visual reiteration of both the lack of government intervention and the resulting extensive hardships for citizens. Such composition testifies to exclusion from formal socio-economic participation and embeddedness in informal economies – the latter inferred by the shot's mise-en-scène of old buildings and spontaneous makeshift roadside businesses. Instead of a progressive life, these micro-economies materialize the frail and temporary livelihoods at the city's edge, facets of the city's "intensifying and broadening impoverishment and rampant informality" (Simone 2001, 16). The street micro-economies represented by these roadside enterprises are off-grid, individual-based survival entrepreneurship. One may aver that they are symbols of how Kinshasa's residents "circumvent the state" (De Villers and Tshonda 2004, 141), by assigning themselves portions of the city outside the mainstream capitalist economic practices common among Kinshasa's kleptocracy. Kristien Geenen (2009, 349) terms such practices a "quiet and seemingly non-organized character of citizens' incursions … the slow occupation of public spaces by the characters, a movement that is irreversible … very powerful."

The wide-angle objective morning shot showing the movement of the crowds away from the modern city towards its margins is an equally useful cinematic device in that it suggests a transition from a town-centred view of Kinshasa to an edge-centred one. In this cinematography, the film responds to an important conversation that has occupied post-independence scholarship about Kinshasa: that formal Kinshasa in the background, once the administrative and geographic centre, has "long since ceased to be either and ... [has] become peripheral to the daily experience of the majority of Kinshasa's population" (De Boeck and Plissart, 2004, 33). By this reasoning, the image summons Kinshasa's historical political logic of detached affiliations between state and citizens. For Mobutu Sese Seko, Kinshasa was largely a hinterland to his Gbadolite jungle palace, with only a neighbourly presence punctuated by his occasional patronage aboard the Kamanyola yacht moored on the Congo River. In this arrangement, while elites wined and dined aboard the yacht, Kinshasa's residents experienced an extreme humanitarian crisis (Ngolet 2011). For Laurent Kabila, Kinshasa was always a city at the edge that he never fully tamed. Despite DRC's optimism at the end of Mobutu's self-serving rule in 1997, recent media reports have referred to the immediate former incumbent political leader as Congo's New Mobutu (Bavier 2010). In this history, we can sense the contiguity of economic plunder. It is in this context, then, that we can read the morning "migration" away from postcolonial Kinshasa as potentially exemplifying a resignation to turmoil in everyday survival, prolonged poverty, and illiteracy (Iyenda 2005). The edge – where the crowds are heading and where most of the film is set – thus embodies the catastrophes of perpetual *laissez faire* economy.

The high-angle camera placement and the elevated gaze that it signifies confer upon the crowds an ambiance of powerlessness, oppression, and defeat. While this may be indicative of an overpowering grip of fate, it stirs ideas of patronage and humiliation as a possible consequence of the characters' detachment from mainstream economic practices. This detachment is achieved using linear perspective and composition of space showing the distance between the modern city in the background and its edge in the foreground. The mise-en-scène of squalid infrastructures along the street's edge, the tedium implicit in the walking masses, and the carceral nature of these businesses add to this aura of individualized survival. If we consider that the long-shot framing renders the scene without any focal point, then a visual reading which gives all these elements equal attention becomes obligatory. Hence, in totality, the opening shot conveys despair in almost all aspects of urban life in Kinshasa. Yet, at close inspection, it is noticeable that what appears an attempt to capture an authentic image of ordinary life at Kinshasa's edge indeed renders a delicate portrait of urban dystopia embalmed in the form of severe sacrifice. The masses render the verisimilitude of doing without public mobility and without good roads. The roadside businesses exhibit frail economic survival, augmenting the shabby rusty houses surrounding the streets to conjure up notions of distraught citizenship. This correspondence between mise-en-scène elements, cinematography, and composition in the opening scene demonstrates the decadence of individual economic enterprise and urban poverty as an inalienable outcome of uncontrolled market practices.

This sense of distress becomes clear in a subsequent scene where some characters push along the road a stalled van with passengers still inside. The blue and yellow of the van's side panels have an indexical relation to the colours of the DRC flag – signifying peace and wealth and a radiant future – making the van a metonymic symbol of the nation. Accordingly, the immobile van reliant on human effort to propel it along the streets becomes a metaphor of a stalled nation. Per this reading, the van reproduces economic hierarchies in the form of master–servant, insider–outsider associations. That the characters burdened with "pushing" the nation forward are "left out" of its ideals of "comfort" reifies an exclusion indexical of Kinshasa's legacy of barbarous economic opportunism and political, social, and economic predation. The interaction between the characters and the van in this shot discloses the severe shortages made possible through the opportunistic economic relations that feature in the film.

It is with this background of economic destitution, exploitative relations, individualistic enterprises, and acute fuel shortages that we interrogate Riva's arrival in Kinshasa with boatloads of fuel, later in this sequence. From a montage perspective, the prioritization of urban distress prior to introducing Riva underscores the latter's opportunistic mentality. His supporting character, G. O., who is his business associate, adds to this sense of economic opportunism. He advises Riva to hoard the fuel and sell it when prices soar due to the acute shortage. Subsequent hoarding of fuel at a moment of crisis to speculate for higher prices expresses the centrality of liberal economic relationships and individual profiteering in the film. Further, that Riva inhabits dingy spaces and brothels despite possessing huge quantities of fuel illustrates the survivalist nature of street economies. He is a prototype for an urbanism that cannot improve one's living standards but instead thrives on "exploitation and growing forms of differentiation" (Freund 2007, 162), with differentiation here anticipating the widening social and economic stratifications. Through such characterization, Riva becomes the quintessential postcolonial Kinshasa's urban dissident: excluded from the benefits expected in the postcolonial city and judged as an alien of sorts (Farber 1996, 84). If we see his squalor as an inventory of urban struggles within the context of the liberalized economy and his fuel enterprise as a *laissez faire* economic practice, then his experiences become a useful lens through which to understand how dystopia works as a survival infrastructure in postcolonial Kinshasa.

Economies of dystopia

Oppong and Woodruff (2007, 10) paint a rather grim picture of the DRC's economy. Calling it "a country of extreme paradoxes," these scholars decry the cyclic suffering, plunder, violence, and poverty that prevail despite the country's vast mineral reserves. These contradictions are more intense in the city of Kinshasa, where exploitation and plunder are covertly embedded in everyday manoeuvres. Trefon (2002, 485) even suggests that Kinshasa is transforming from "*kin la belle* into *kin la poubelle*" ("Kinshasa the beautiful, Kinshasa the dump"), alluding to the city's recursive decadence. The paradox lies in DRC's elaborate history of crises manifesting in, among others, contiguous political and economic turmoil (Nzongola-Ntalaja 2002), despite significant effort to correct the scenario. An effort to cast off what they perceived as "negative baggage" (Dunn 2003, 110), Mobutu's 1971 and Laurent Kabila's 1997 political decisions signalled a "re-baptism" of the country's postcolonial consciousness. Mobutu's authenticity initiative which sought to restore authentic African identity (Adelman 1975; Dunn 2003) to the nation and Laurent Kabila's reroute strategies which sought to reinstate national ideals can be read in two ways: either as attempts to create a universally popular symbol of postcolonial DRC, where distribution of benefits from the country's vast mineral economy was the expectation, or as impromptu responses from the respective leaders when citizens encountered severe economic distress for which the citizens were not prepared. Given that both attempts failed, many residents of Kinshasa found themselves retreating even further into already normalized informal identities that existed in response to the city's persistent economic crises.

In the public urban form symbolized by the street life in Kinshasa, dystopian existence is more than a condition. It exhibits the inner logic of how those trapped in these conditions make, do, and survive by adapting themselves accordingly. The Khadaffi and la sapeur, two prominent street identities in Kinshasa, imbricate two responses to urban dystopia that have been practiced in the city, namely, an aggressive practice of profiteering and dreamy impressions of opulence. Street vendors specializing in fuel, Khadaffi represent les Lutteurs – Kinshasa's street strugglers. In contrast, la sapeur is a bashful character who dresses smartly to create an impression of opulence. Amid acutely insufficient economic resources, the hazardous lifestyles embodied by these two characters have become inventive infrastructures of survival. I call these "survival infrastructure economies of dystopia."

The use of the term dystopia here is contrary to the ideas of urban dystopia based on "the privatization of the city with the deepening of urban segregation and social-spatial exclusion" (Pow 2015, 464). For, although such installations are critical markers of urban divisions, they offer little insight into the survival modes that such divisions make possible and inform us even less about how crisis becomes an asset to those who cannot escape it. Thus, although Kinshasa is highly segregated and a place where inequality thrives, the starting point for studying such urban configurations must not always begin with the formal city as the mainstream, and the informal as the "other." Such a common-sense reading of Kinshasa, where informality has largely become the main mode of urbanism, may not fully address the city's present realities (Baeten 2002, 103). The approach to dystopia adopted here seeks to "accept deviance, conflict, difference and oppositional interests as (potentially) creative rather than destructive forces in today's city" (Ibid., 103–104).

The discussion of urban dystopia contemplated here, then, is not so much about social, economic, or political difficulties; it is about the forms of urbanism resulting from these difficulties.

The concept of economies of dystopia, then, is about the inability to tap into networks requisite for collective survival. In response, one undertakes individualistic decisions to cope day by day, confronting problems as they arise here and there. To live a life of dystopia is to live as a pariah: not necessarily "at the bottom of the social or economic ladder," but to be loathed, shunned, and isolated (Farber 1996, 266). By sidestepping the mainstream city and emphasizing the street life in Kinshasa's periphery where economic degeneration is more pronounced, *Viva Riva!* emphasizes certain characteristics of Kinshasa residents who use crises as an infrastructure for survival. How do we make sense of such fragile survival, enclosed between freedom and despair, within Kinshasa's postcolonial urbanism? In *Viva Riva!*, the Khadaffi's aggressive profiteering depends on a fuel crisis; his pleasures, mostly wanton sexual indulgences, occur in the context of potential health hazards. Comparably, although la sapeur exhibits restraint and commands social power through his impressive self-presentation, his financial crisis is a major motivation behind his curated appearances. In these considerations, Riva and Azor, two key characters in *Viva Riva!*, exemplify Khadaffi and la sapeur, respectively. I discuss these two framings of dystopia to illustrate how *laissez faire* economic practices in postcolonial Kinshasa not only characterize Kinshasa in this film but also offer a perspective on how dystopian economic conditions have bred new models of urbanism.

Khadaffi

Informal street trade is common in many cities across DRC. In most of Kinshasa, street vending exists outside

> All legal trade and economic regulations (i.e. no license, no insurance, no minimum wage, no health and safety standards) and bureaucratic rules. Goods are sold from fixed stalls, from the pavement, in front of people's houses and doors, in small shops and, in most cases, from people's heads as they walk along the streets
>
> (Iyenda 2005, 58)

That such abstract practices have become the norm for many Kinois suggests informal economic practices as thriving responses to the rampant unemployment and poverty that preoccupy the majority of Kinshasa residents (Iyenda 2005, 55). That is, rather than indexing economic mobility or opportunity, they are manifestations of a desperate response to economic crisis, day after day. As the city's fuel supply continues to be erratic, one of the most notable street vendors is the fuel trader, the Khadaffi. The name Khadaffi is derived from Colonel Gaddafi of Libya, known for his involvement in the fuel business in later years (Lamarque 2014). In DRC, Khadaffi refers to the cunning businesspeople who hoard fuel and avail it at moments of scarcity (Bilakila 2004, 28). Dominating the streets in such moments, the Khadaffi's profiteering from the fuel business gives them an economic upper hand to undercut the city's mainstream fuel businesses. The first framing of dystopia in *Viva Riva!* is built around this street-smart fuel entrepreneur.

Early in *Viva Riva!*, we encounter scenes of an acute fuel shortage in Kinshasa. Starting with the stalled van and continuing in subsequent scenes where we see lines of cars queuing at a petrol station and others refuelling from roadside fuel vendors, the film's mise-en-scène, composition, and montage indicates severe fuel shortage. The shot of the petrol station includes a prominent handwritten sign reading "Plus de Carburant" (translatable to "More Fuel") and a long queue of cars waiting to refuel, illustrating the severity of the fuel crisis. It is within this context that we can read a subsequent montage, whose first shot, framed as a close-up, shows an informal fuel vendor fuelling a car using an improvised bottle funnel. The mise-en-scène of this shot comprises fuel in a jerry can, a funnel improvised from a plastic bottle, and the fuel pouring into a car's fuel tank. The shot truncates the character almost entirely, giving prominence to the fuelling process. Using this framing, composition, and mise-en-scène, the shot posits the improvised "fuelling station" as the main element in the shot – making Khadaffi enterprise a critical urban practice in this film. These scenes provide a background to Riva's arrival in Kinshasa with boatloads of fuel. Through this montage, the film has primed the viewer to read Riva's arrival in Kinshasa with fuel barrels at a moment of extreme scarcity as Kinshasa's ultimate Khadaffi. Throughout this film, Riva's business of hoarding and selling fuel is central both to his character development and the narrative progression.

The second scene, showing a character siphoning fuel from a car, also starts with a close-up shot where we see a jerrycan, a piece of hosepipe, and a character siphoning fuel from the car's fuel tank. The scene emphasizes both the process of siphoning and the character involved, implicating the character in the undoing of the fuelling undertaken in previous shot. The Khadaffi fuelling and the one siphoning are partaking in a cyclic network that does not grant either of them progress but guarantees survival in the city (Geenen 2009, 350). They recycle the same problem of fuel scarcity by first appearing to solve it by fuelling, only to cause it again by siphoning. The deal made with the Khadaffi who fuels the car is immediately revoked when the other Khadaffi siphons the same fuel.

In this dialectic of fuelling and siphoning, we see a "protracted struggle over the legitimacy of self-employment and of the right to survive in the city" (Simone 2004, 169). Through such deceitful means, city residents partake in a performance of cyclic crisis, where individual economic interests come before business ethics or collective progress. This cycle exemplifies how *laissez faire* economic conditions, evident in the commodification of the fuel crisis, motivates exploitative urban practices in the city whereby individual characters betray and exploit each other for small gains, unwittingly getting closer to their eventual fatality. The kind of urbanism actuated through the

Khadaffi character, and on which the character of Riva is built, embraces the risk of ever-present fatality. In *Viva Riva!,* characters' betrayals and fatalities are central to their urban experiences. When the policewoman shows Cesar's gang the lorry that transported Riva's stolen fuel to Mariano, she is betraying Paul (Richard Manda), the driver who is then fatally assaulted by Cesar's gang. Père Gaston (Bavon Diana Landa), a church leader, charges Cesar's gang an exorbitant price to hide them as fugitives, going on to betray his church by collecting all their offerings and trying to purchase fuel with it. In the end, he is killed (betrayed) by Cesar's gang. Commissaire de Police (Elbas Manuana) in charge of Lwano police station asks for a USD $1,000 bribe to release the gang, betraying his country by compromising the duties assigned to him as the guarantor of civilian security. He is also killed by Cesar's gang. The same template is evident in the character arcs of G. O., J. M., and Malou – all of who betray Riva to Cesar's gang during the film. By this reasoning, Khadaffi can be interpreted as a mentality that exhibits the reality of non-progressive life in Kinshasa. Through this character, the film makes overt connections between fuel entrepreneurship and historic economic exploitation, thus inviting a discussion of the scene of Riva arriving with boatloads of fuel, which characterize him as Khadaffi *par excellence*, as the main motif through which this film communicates the centrality of crisis in Kinshasa's economic life.

Such profiteering, constitutive of Riva's moniker of paltry survivor, trickster, and degenerate character, aptly invokes De Boeck's (2004, 46) concept of *nzombo le soir*, an expression standing for "an unexpected opportunity or windfall within an overall condition of lack and want." His material lifestyle, animated through consumption in night clubs and moments of reckless pleasure in brothels, is built around the windfall profits from his fuel business. Possessing fuel is thus irreducible to the mere act of hoarding and speculative entrepreneurship; it exemplifies the logic of *Kobeta libanga*, implying "trickery, 'wheeling and dealing', acting as a go-between or bargaining" (Bilakila 2004, 20). The Khadaffi's urban behaviour follows a well-established *laissez faire* economic practice, putting the individual's enterprising acumen and the tragedies that it produces at the centre of urban culture. Such practices, the film suggests through the fatalities of its characters, create a perilous urban experience whereby each moment is hazardous.

Khadaffi's construal of urban dystopia can be read as a metaphor of life and death:

> How then, does one live when the time to die has passed, when it is even forbidden to be alive, in what might be called an experience of living the "wrong way round"? How, in such circumstances, does one experience not only the everyday but the *hic et nunc* when, every day, one has both to expect anything and to live in expectation of something that has not yet been realized, is delaying being realized, is constantly unaccomplished and elusive?

> (Mbembe 2001, 201)

In juxtaposing life and death, Mbembe hybridizes the condition of hazardous existence so that the urban subject exists in perpetual risk of danger. The dominant force in urban existence is thus not one's plans but the unforeseen dangers through which those plans are actuated. Take, for instance, Riva's character arc as a perpetual street hustler. Despite having economic resources (fuel) and being characterized as Kinshasa's biggest Khadaffi, Riva lives his life the "wrong way round." Not only does he live speculating on fuel prices, but he is also, every day, speculating on his life. Through a network of informers, including brothel mistresses and the street boy Anto, he is engaged in a dangerous struggle with an Angolan gang, and also with the city. Throughout the film, he is a fugitive displaced from the freedom he seeks. The Kinshasa of prosperity that Riva came to settle in – the one he seeks throughout the film – is constantly threatened by the dangerous circumstances he has to endure each day: a fugitive from a ruthless criminal gang, betrayal from close friends, excommunication from family, and the danger of Azor's revenge. Thus, instead of Riva embodying Kinshasa's economic potential and peaceful habitation of the city, the character of the Khadaffi in this film does not question values, morals, or ethics but aids an understanding of the "evolving material interrelations ... 'metabolic relations' ... between human beings and nature" (Swyngedouw 2006, 25). Each character connected to the Khadaffi economy is exposed to some sort of danger, whether disease, poverty, violence, or exploitation. Yet the Khadaffi thrives amidst such dystopia.

La sapeur

In DRC, sapeurism is a dominant urban culture with roots both in colonial Kinshasa and across the Congo River, in Brazzaville (Republic of the Congo). In postcolonial DRC, "*la sape*," a short form of La Société des Ambianceurs et des Personnes Élégantes (the Society of Tastemakers and Elegant Persons), still refers to a person who dresses fashionably to mimic an identity of whiteness (Friedman 1992) or to assert new cultural sensibilities and an accumulation of social status (Finkelstein 1997). The sapeurs are also associated with the use of elegance to borrow new identities (Gondola 1999). La sapeur is thus a dreamy urban character who uses fashion as escapism from urban pressures. Here, after natives were paid with clothes for their work, they adopted flashy dressing to "combat inferiority leveled at them by their French and Belgian masters and ... adapt to their master's style but with a hint of their very own exaggerated high-fashion style" (Douniama 2018). Sapeurism substituted legitimate progress and meaningful economic reward with impressions and voyeurism. La sapeur's impressions of economic achievement and idealistic self-presentation are a therapeutic exercise in a dystopian urban condition. In Kinshasa, la sapeur is contrasted with les Lutteurs (the strugglers), the aggressive street hustlers unashamed to show their hardships. The la sapeur character thus offers a unique perspective into passive responses to Kinshasa's difficult conditions. This subsection discusses the representations of this character in *Viva Riva!* as the second framing of dystopia in the context of *laissez faire* economic practices in postcolonial Kinshasa.

In postwar DRC, les sapeurs play several roles. On a personal level, they embody the revival of the ideals of "respect, peace, integrity, and honour … a sapeur has to be non-violent, well-mannered, and inspiration through their attitude and behaviour" (Lyons 2014). Given that these attributes and experiences are outside the daily reach of many residents of Kinshasa, la sapeur fulfils a voyeuristic ideal for many Kinois. He represents hope within bleakness, his performance of affluence actualizing if only at a cathartic level, a spectacle of popular illusions. Kinshasa's les sapeurs are thus placeholders of an imaginary urban condition (of opulence, peace, and progressive culture). They contrast with the street hustlers who embody a verisimilitude of struggle, dishonour, and retrogression. In Kinshasa, this self-representation is a counter-narrative aimed at exchanging the city's dystopic image with a positive imaginary acceptable in the eyes of the city's residents (Lanquetin 2010). At a national level, la sapeur is a cultural icon, especially of the Rhumba music popular among Congolese. He embodies both the commercialization of elegance and pursuit of inclusive urban community. The sapeur culture has enabled a classless community to flourish in a highly classed city. They are thus emblematic of urban tranquillity. At a public level, la sapeur is also a national figure who opens practices of "economics of transnationalism … [by] flaunting their wealth … following a frugal period of hard work" (Eriksen 2007, 94). Their existence in Kinshasa parades the benefits of *laissez faire* economic practices freely for all to see; their lifestyles readily becoming performances of the socially affable, self-regulating, wealthy individual (Smith 1776). In a discussion of his *Pièces d'identités* (1997), Ngangura Mweze describes les sapeurs as individuals "who have no other objective in life than to have nice clothes and can even commit crimes to sustain their chosen lifestyles" (Ukadike 2002, 145). La sapeur is thus a holographic sign of how Kinshasa residents fumble with life, barely surviving, while postponing indefinitely the inevitable consequences of the city's dystopia. Based on this concept of la sapeur in postcolonial Kinshasa, one may argue that *Viva Riva!'s* use of this character promotes a subtext of passive response to urban dystopia in Kinshasa. In this film, Azor characterizes la sapeur.

The ultimate la sapeur in *Viva Riva!*, Azor, is also one of the more well-developed characters in this film. He is introduced at the same time Riva enters Kinshasa's upper-class social world and remains central to the film's narrative of economic tragedies. Azor's delayed entry into the narrative plays a crucial role in distinguishing him from les lutteurs, who are embodied by all the other characters. The initial sequence of the film is about urban suffering and unproductive street entrepreneurship. Thus, isolating Azor from this display of squalor plays a key role in grounding his identity in the city. While Riva and Cesar chase each other in the streets over control of the fuel economy, Azor lives in his exclusive mansion, always attended by his bodyguards, and accompanied by Nora, a beautiful young woman. Unlike all the other characters in this film, Azor is seen as the ultimate symbol of urban success, and Kinshasa's social community – represented by nightclub patrons – does not hesitate to notice and even applaud this impression.

The first time we see Azor is inside the Sai Sai nightclub, where he is introduced as "the big boss, the descendant of all the Kings of the Congo, the strong man of Kinshasa. Azor the first." Amid applause from the club patrons and the club owner, Azor casts a refined figure of opulence. No other character in this film matches his grandeur, self-restraint, and finesse of etiquette. To fully characterize his social standing, the film uses such endorsement to acknowledge and grant him privileged access to Kinshasa's upper circles. Even his residence is set apart from the rest of the city: unlike the squalid, congested residents where Riva, J. M., their families, and the prostitutes at Mother Edo reside, Azor's home is in a walled compound at the top of the hill. He also has sophisticated tastes: the poster of the infamous 1974 Rumble in The Jungle fight between Mohammed Ali and George Foreman marks him as an art collector, while wine and cars are props imbuing him with social power. Compared to Riva, whose sexual activities with prostitutes are framed to emphasize the disgust of the naked bodies painted with colours and adorning masks, Azor has a beautiful girlfriend. This social contrast between Azor and Riva is an important signifier that points us towards the centrality of performed success in Kinshasa's everyday urbanism.

It is thus notable when Azor refers to Riva as a peasant during their first altercation inside Sai Sai nightclub. This proclamation affirms economic boundaries as simultaneously social boundaries and becomes a fulcrum for debating urban identities in postcolonial Kinshasa. It positions Azor – and hence sapeurism – at the centre of this debate. Equally notable is the subsequent characterization of Azor as economically struggling and sexually incapable of satisfying his girlfriend. When we see Nora confronting Azor over sexual dissatisfaction and finally indulging in sexual relations with Riva in Azor's home before eloping to be with him, we start to see Azor's inadequate sexual vigour as a placeholder for other serious weaknesses. This is amplified and ridiculed by juxtaposing his behaviour with Riva's insatiable sexual appetites, which include activities with prostitutes and luring Nora from Azor. This outcome in Azor's personal life is mimicked in his financial life when we see him nicking Nora's earring to pay for his bills at Sai Sai nightclub after the proprietor threatens to evict him and his bodyguards – a scene juxtaposed with one of Riva buying beer for all the Sai Sai patrons just after Azor has nipped Nora's earring for lack of money. Such juxtapositions portray la sapeur as a peddler of impressions rather than of real power.

Consequent to this framing of Azor's contradictory character – endowed with immense social power yet imbued with severe deficiencies – his esteemed status of the "strong man of Kinshasa" draws attention to the way the film constructs the quandary of dystopic existence. Azor functions as an exposé of erratic "complexities of survival 'here and now' and uncertainty about the future" (Bilakila 2004, 23), meaning his elegance and low-key elitism indeed mark him more as a la lutteur. That his initial iconicity of opulence is reduced to a symbol of doing without casts him as a palimpsest of Kinshasa's postcolonial destitution, what Mbembe (2001, 201) likens to "live in death," or an "appearance" (Ibid., 186). Mbembe's idea of the postcolonial subject as an appearance describes not only the continuing de-personalization of the African by the former colonialists but also a certain level of realism in the postcolonial urban condition, herein termed *laissez faire* urbanism. Wa Munga's character-ization of Azor specifically, and the rest of the characters inhabiting Kinshasa's edges in general, carries with it this consciousness of the urban citizen tightly enmeshed in the global networks of capitalist hegemony, yet inextricably pro-pelled by a delusional obsession with making impressions.

We can thus critique la sapeur through Mbembe's idea that "in Africa after colonization, it is possible to delegate one's death while simultaneously and already experiencing death at the very heart of one's existence" (Mbembe 2001, 201). Azor's character is not about the verisimilitude of daily struggles in Kinshasa but about displaying the concurrence of abundance and destitution, riches and poverty, good and evil as an onerous legacy of unfruitful urban enterprises (Iyenda 2005, 65). La sapeur is not just a fictitious enabler of appear-ances, it is an infrastructure to survive Kinshasa's dystopic conditions. Sapeurism, like Khadaffi, is a narrative device that alerts the viewer to the uni-versality of *bashege*, the social rejects who thrive off informality (De Boeck 2004). Bashege here is a Kinois' term for Kinshasa's street children and is meta-phorical of extreme economic difficulties. Like Riva, Azor is acutely involved in *"débrouillez-vous"* urban citizenship, that is, confronting scarcity by instanta-neously figuring out survival possibilities where and when they become neces-sary (Trefon 2002, 488). Consequently, through Riva and Azor, the film characterizes Kinshasa's urbanism as a "pathological milieu" (Press and Smith 1980, 4), a city whose residents use appearances to enfold the various forms of dystopia. While Riva and Azor's characters appear contradictory and exist in perpetual conflict within the film, this conflict does not "determine who will have the freedom of the streets and who will not" (Wood 2012, 51), but accentu-ates la sapeur's and la lutteur's equanimous responses to Kinshasa's carceral conditions, whether in the form of charismatic social mobilization as in Azor's case, or aggressive economic mobilization as used by Riva. That both characters are struggling against the city rather than against each other becomes clear in the film's tragic ending where they all die. The death of all prominent characters at the end of *Viva Riva!* challenges the viewer to question who are the film's real protagonist and antagonist. It is reasonable to argue that the film posits Kinshasa's *laissez faire* economic practices, symbolized by fuel as the antago-nist, while the characters entrapped in the disastrous consequences of such an unrewarding economic situation are the protagonists. Their decimation, all tied to the pursuit of economic betterment, alert us to the undeniable role of dysto-pia in Kinshasa's everyday practices. In the ensuing scenario of murders and betrayals and risky lifestyles and poverty, dystopia (a consequence of *laissez faire* economic practices) greatly determines how the characters inhabit the city.

Afterthought

In the early scenes of *Viva Riva!*'s final sequence, Kinshasa is likened to muddy garbage. Just before we see the image of the warehouse where Riva's smuggled fuel is hoarded, the director indulges the audience with an image of Cesar leading a La Commandante and J. M. as captives through a street full of mud and garbage. A voice-over announces: "Kinshasa the beauty, Kinshasa the garbage," creating an aural and visual harmony with the shots of a La Commandante walking on the muddy junk. The camera persists with the same close-up shot framing as it tilts up tracing her body from her feet stepping on the mud to her face. The persistent close-up framing and tilt-up camera movement focuses the viewer's attention on her frailty in the moment of her greatest despair. As we watch the once-powerful military officer hustled by a foreign gang, it becomes obvious that the pursuit of individual profiteering in Kinshasa has prevailed over personal freedom. Cesar's comment at the end of this shot: "Your country is the worst cow pee I have ever seen, maybe you should have remained colonized," communicates the way individual economic interests in Kinshasa override collective wellbeing. Hence, the film's fatal ending whereby all characters die and the fuel is set ablaze, which is precipitated by ungoverned economic interests catastrophizing profiteering in postcolonial Kinshasa, offers a reasoned imaginary of Kinshasa's informal economies as prevalently dystopic. How this reasoning could be interpolated in DRC's political economy as a basis for further studies of the city remains a fascinating route.

Bibliography

Adelman, Kenneth Lee. 1975. "The Recourse to Authenticity and Negritude in Zaire." *The Journal of Modern African Studies* 13 (1): 134–139.

Ayimpam, Sylvie. 2019. "Street smarts in Kinshasa." *The UNESCO Courier - April-June:* 12–14.

Baeten, Guy. 2002. "Hypochondriac Geographies of the City and the New Urban Dystopia." *City* 6 (1): 103–115.

Bavier, Joe. 2010. "Congo's New Mobutu." *Foreign Policy,* June 29, 2010. https://foreignpolicy.com/2010/06/29/congos-new-mobutu/.

Bazenguissa-Ganga, Rémy. 2014. "Beautifying Brazzaville: Arts of Citizenship in the Congo." In *The Arts of Citizenship in African Cities: Infrastructures and Spaces of Belonging*, edited by Mamadou Diouf and Rosalind Fredericks, 163–186. New York: Palgrave Macmillan.

Bilakila, Anastase Nzeza. 2004. "The Kinshasa Bargain." In *Reinventing Order in the Congo: How People Respond to State Failure in Kinshasa*, edited by Theodore Trefon, 20–32. London and New York: Zed Books.

Bouwer, Karen. 2019. "Life in Cinematic Urban Africa: Inertia, Suspension, Flow." In *A Companion to African Cinema*, edited by Kenneth W. Harrow and Carmela Garritano, 69–88. Hoboken: John Wiley & Sons.

Chang, Chris. 2011. "Short Takes: Viva Riva!" *Film Comment,* May/June. https://wwww.filmcomment.com/article/viva-riva-review/

De Boeck, Filip. 2004. "On Being Shege in Kinshasa: Children, the Occult and the Street." In *Reinventing Order in the Congo: How People Respond to State Failure in Kinshasa*, edited by T. Trefon, 155–173. London and New York: Zed Books.

De Boeck, Filip. 2011. "Spectral Kinshasa: Building the City through an Architecture of Words." In *Urban Theory Beyond the West: A World of Cities*, edited by Edensor Tim and Mark Jayne, 309–326. London and New York: Routledge.

De Boeck, Filip, and Marie-Francoise Plissart. 2004. *Kinshasa: Tales of the Invisible City*. Leuven: Leuven University Press.

De Herdt, Tom, and Wim Marivoet. 2018. "Is Informalization Equalizing? Evidence from Kinshasa (DRC)." *Journal of Contemporary African Studies* 36 (1): 121–142.

De Villers, Gauthier, and Jean Omasombo Tshonda. 2004. "When Kinois Take to the Streets." In *Reinventing Order in the Congo: How People Respond to State Failure in Kinshasa*, edited by Theodore Trefon, 137–154. London and New York: Zed Books.

Douniama, Victoire. 2018. "The Art of La Sape: Fashion Tips from Congo's 'Sapeurs'." *The Culture Trip*, June 25, 2018. https://theculturetrip.com/africa/congo/articles/the-art-of-la-sape-fashion-tips-from-congos-sapeurs/.

Dunn, Kevin C. 2003. *Imagining the Congo: The International Relations of Identity*. New York: Palgrave Macmillan.

Eriksen, Thomas Hylland. 2007. *Globalization: The Key Concepts*. Oxford and New York: Berg.

Farber, Daniel A. 1996. "The Pariah Principle." *Constitutional Commentary* 13: 257–284.

Finkelstein, Joanne. 1997. "Chic Theory." *Australian Humanities Review*, March 1. http://australianhumanitiesreview.org/1997/03/01/chic-theory/

Frassinelli, Pier Paolo. 2015. "Heading South: Theory, Viva Riva! and District 9." *Critical Arts* 29 (3): 293–309.

Freund, Bill. 2007. *The African City: A History*. Cambridge: Cambridge University Press.

Friedman, Jonathan. 1992. "Narcissism, Roots and Postmodernity: The Constitution of Selfhood in the Global Crisis." In *Modernity and Identity*, edited by Scott Lash and Friedman Jonathan, 167. Fricdman. Oxford: Blackwell

Geenen, Kristien. 2009. "'Sleep Occupies No Space': The Use of Public Space by Street Gangs in Kinshasa." *Africa: Journal of the International African Institute* 79 (3): 347–368.

Gondola, Didier. 1999. "Dream and Drama: The Search for Elegance among Congolese Youth." *African Studies Review* 42 (1): 23–48.

Iyenda, Guillaume. 2005. "Street Enterprises, Urban Livelihoods and Poverty in Kinshasa." *Environment & Urbanization* 17 (2) (October): 55–67.

Lamarque, Hugh. 2014. "Fuelling the Borderland: Power and Petrol in Goma and Gisenyi." *Articulo: Journal of Urban Research* 10. https://journals.openedition.org/articulo/2540.

Lanquetin, Jean-Christophe. 2010. "Sape Project: 2006–2009." In *African Cities Reader*, edited by Ntone Edjabe and Edgar Pieterse, translated by Dominique Malaquais, 41–49. Vlaeberg: Chimurenga Magazine and the African Centre for Cities.

Lyons, Juliette. 2014. "La Sape: An Elegance that Brought Peace in the Midst of Congolese Chaos." *Le Journal International*, May 12. https://www.lejournalinterna-tional.fr/La-Sape-an-elegance-that-brought-peace-in-the-midst-of-Congolese-chaos_a1871.html.

Mbembe, Achille. 2001. *On the Postcolony*. Berkeley/Los Angeles/London: University of California Press.

Neuwirth, Robert. 2005. *Shadow Cities: A Billion Squatters, a New Urban World*. London and New York: Routledge.

Ngolet, François. 2011. *Crisis in the Congo: The Rise and Fall of Laurent Kabila*. New York: Palgrave Macmillan.

Nzongola-Ntalaja, Georges. 2002. *The Congo: From Leopold to Kabila: A People's History*. London and New York: Zed Books.

Oppong, Joseph R., and Tanya Woodruff. 2007. *Democratic Republic of the Congo*, Edited by Charles F. Gritzner. New York: Infobase Publishing.

Pieterse, Edgar. 2011. "Grasping the Unknowable: Coming to Grips with African Urbanisms." *Social Dynamics: A Journal of African Studies* 37 (1): 5–23.

Pow, Choon-Piew. 2015. "Urban Dystopia and Epistemologies of Hope." *Progress in Human Geography* 39 (4): 464–485.

Press, Irwin, and Estellie M. Smith. 1980. *Urban Place and Process: Readings in the Anthropology of Cities*. New York: Macmillan Publishing Co., Inc.

Pype, Katrien. 2007. "Fighting Boys, Strong Men, and Gorillas: Notes Ont He Imaginations of Masculinites in Kinshasa." *Africa: Journal of the International African Institute* 77 (2): 250–271.

Pype, Katrien. 2012. *The Making of the Pentecostal Melodrama: Religion, Media, and Gender in Kinshasa*. New York and Oxford: Berghahn Books.

Simone, AbdouMaliq. 2001. "On the Worlding of African Cities." *African Studies Review* 44 (2) Ways of Seeing: Beyond the New Nativism: 15–41.

Simone, AbdouMaliq. 2004. *For the City Yet to Come: Changing African Life in Four Cities*. Durham and London: Duke University Press.

Simone, AbdouMaliq. 2007a. "Assembling Douala: Imagining Forms of Urban Sociality." In *Urban Imaginaries: Locating the Modern City*, edited by Alev Çınar and Thomas Bender, 79–99. Minneapolis: University of Minnesota Press.

Simone, AbdouMaliq. 2007b. "Deep Into the Night the City Calls as the Blacks Come Home to Roost." *Theory, Culture & Society* 24 (7–8): 235–248.

Simone, AbdouMaliq. 2011. "The Ineligible Majority: Urbanizing the Postcolony in Africa and Southeast Asia." *Geoforum Geoforum* 42: 266–270.

Smith, Adam. 1776. *An Inquiry Into the Nature and Causes of the Wealth of Nations*. London: W. Strahan and T. Cadell.

Swyngedouw, Erik. 2006. "Metabolic Urbanization: The Making of Cyborg Cities." In *In the Nature of Cities: Urban Political Ecology and the Politics of Urban Metabolism*, edited by Nik Heynen, Maria Kaika, and Erik Swyngedouw, 20–39. London and New York: Routledge.

Tchouaffe, Olivier J. 2016. "Perspectives on New Popular African Cinema, History and Creative Destruction in Viva Riva! (2011)." *Journal of African Cinemas* 8 (3) December: 299–312(14).

Trapido, Joe. 2016. *Breaking Rocks: Music, Ideology, and Economic Collapse, From Paris to Kinshasa*. New York and Oxford: Berghahn Books.

Trefon, Theodore. 2002. "The Political Economy of Sacrifice: Kinois & the State." *Review of African Political Economy* 29 (93/94) State Failure in the Congo: Perceptions & Realities (*Le Congo entre Crise et Régénération*): 481–498.

Trefon, Theodore. 2004. "Introduction: Reinventing Order." In *Reinventing Order in the Congo: How People Respond to State Failure in Kinshasa*, edited by Theodore Trefon, 1–19. London and New York: Zed Books.

Ukadike, Nwachukwu Frank. 2002. *Questioning African Cinema: Conversations with Filmmakers*. Minneapolis: University of Minnesota Press.

White, Bob W. 2008. *Rumba Rules: The Politics of Dance Music in Mobutu's Zaire*. Durham: Duke University Press.

Wood, Phil. 2012. "Challenges of Governance in Multi-ethnic Cities." In *In Cities, Cultural Policy and Governance*, edited by Helmut Anheier and Yudhishthir Raj Isar, 44–60. London/Los Angeles/New Delhi/Singapore/Washington DC: Sage Publications Ltd.

Filmography

City of Joy. Dir. Roland Joffé, United Kingdom, 1992.
Congo in Four Acts. Dir. Dieudo Hamadi, Kiripi Katembo, Divita Wa Lusala, South Africa, 2010.
Kinshasa Makambo. Dir. Dieudo Hamadi, The DRC, 2018.
This Is Congo. Dir. Daniel McCabe, USA, 2017.
Viva! Riva. Dir. Djo Tunda Wa Munga, The DRC, 2010.

4 Urbanism of the commons

Inhabiting trash and a crisis of communing in Nairobi

Nairobi

In many aspects, Nairobi city has become a conglomerate of informal habitation. With the city's affluent squeezed in pockets of high-end suburbs surrounded by miles of informal settlements, the city in its current form is mainly informal. Although this structure is a rip-off of urban patterns seen around the continent, it nevertheless has major implications for how the city's space may be theorized in the present. The incapacity to share the city as a common space – a failure that sprouts from contestation over real estate as a scarce urban resource – also produces the rationale that certain urban spaces would be sustainably developed and others, relegated as "trash," would attract insufficient capital investment. This has produced a skyline of dilapidated material structures and modern skyscrapers, both forming a visible frontier between the two cities. In this context, garbage acquires the metonymic value of designating the boundary between the formal and the informal and the legal and the illegal. The operationalization of trash goes beyond the scattering of consumer product wraps and rotting perishable commodities, leaking sewers and chemical effluents, and the various shades of rot and pollution oozing in some parts of the city. Trash here is an allegory of the everyday life of Nairobi residents, who, living lives infiltrated by garbage, inhabit a "trashed city," the only possible site where they can gain inclusion to the city's dwindling spaces and resources. Here, inhabiting trashed spaces expresses a rapidly evolving form of urbanism where trash (and emblematic spaces such as downtown) becomes the only option to claim portions of the city for those with insufficient economic means. Based on this indexicality of trash and the role it plays in designating ways of communing with Nairobi city, this chapter looks at trash's indexicality of urban power tussles, examining how Tosh Gitonga's *Nairobi Half Life* (2012) uses Nairobi's garbage sites to motivate a debate on the urban commons. "Occupation of trash," it is argued, designates a form of crisis urbanism that experiments with tactics around sharing the city as a common space.

Garbage and urban commons

"One glimpse is enough. You have discovered the famous misery of the Third World," states Robert Neuwirth's (2005, 67) introduction to his chapter about

DOI: 10.4324/9781003122098-4

urban squatting in Nairobi's Kibera slums. He proceeds thus: "A sea of homes made from earth and sticks rising from primeval mud-puddle streets ... All, old and young, new arrivals and long-term residents, live without running water, sewers, sanitation, or toilets." This description of human suffering in Nairobi city due to congestion, material deprivation, and lack of social amenities is further compounded by the overwhelming presence of garbage. Neuwirth (2005, 68) describes the trash problem as follows:

Piles of trash line every alley and avenue, giving the neighborhood its trademark look: a motley patina of red dirt, green mango peels, and the festive but faded colors of thousands of discarded plastic bags. Chickens and goats wander by and scratch at the heaps for food. Upon occasion, to reduce the load, someone will rake some of the garbage into a pile, push it to the side or against a mud wall, and set it on fire. These smoldering mounds pose the biggest danger to the community: that the flames will spread to the dry wood of the huts. But what else is there to do with trash? There's no one around to pick it up.

This description of the human experience in Nairobi's Kibera slums, and the position of garbage in accentuating this condition, elucidates the argument pursued in this chapter, that Nairobi city is not shared as a common space, that this lack constitutes a crisis of the city's postcolonial urbanism, and that the presence (and image) of "the trash" is central to explicating this discussion.

Improvising salvaged pieces of junk offers, in the words of Gillian Whiteley (2011, xii), a "perfect analogy for the haphazard and serendipitous nature of one's journey through [city] life." Sometimes the reconstituted garbage teases out underlying urban tensions, as was the case when Joseph-Francis Sumégné installed *La Nouvelle Liberte*, a sculpture modelled from salvaged urban junk, at the Deido roundabout in Douala city, Cameroon (Simone 2004). For Douala residents, the sculpture stirred clashing perspectives. On one hand, city elites felt that "car parts, discarded tires, scrap metals, and other various debris" gave "hypervisibility" to "aspects of the city they had dodged" (Simone 2004, 114). On the other hand, the city's junkies recognized the dregs as an affirmation of "a [recycled] way of urban life that people had been reluctant, even embarrassed, to affirm" (Ibid., 113–114). In this assembled junk form, Sumégné's sculpture retrieved minuscule "traces of [the city's] authentic experience" (Whiteley 2011, 35), connecting stories from within the city's less visible informal spaces to those from its visible formal spaces. Here, junk represents a conflation of seemingly mutually exclusive urban worlds: the affluent formal city and the decrepit informal slums. These clashing views posit junk as a useful metaphor. That publicizing junk represents a re-insertion of that which is rejected and its decrepit subsistence back into circulation as was the case with the *La Nouvelle Liberte* sculpture, means that its existence – even in reconstituted form – represents a "strategic redemption of the low, the despised, the imperfect, and the 'trashy' as part of a social over-turning" (Stam 2003, 35). By using junk to engage with critical issues of urban inequality and politics of belonging unequally in the city, these scholars curate a useful metaphor through which we can engage questions about the urban commons.

Rosalind Fredericks (2018) critiques junk as a metaphor for urban communal activism. She talks about two urban metaphors of garbage: dumping and cleaning. In the former, she writes about a moment when Dakar experienced near-spontaneous trash revolts:

> Its municipal trash workers went on strike and ordinary Dakarois, in solidarity, staged dramatic neighbourhood-wide trash 'revolts' through dumping their household waste into the public space. Across the city, mountains of trash choked the capital's grand boulevards and paralyzed many of the city's functions
>
> (Fredericks 2018, 1)

Although the dumping was initially catalysed by dissatisfaction with Senegal President Abdoulaye Wade's second re-election in 2007, garbage soon came to signify other aspects of the city's management: "trash workers' labour dispute ... the burdens of managing their festering garbage ... the resolution of the material inequalities of urban infrastructure" (Ibid.). Residents blocked Boulevard Dial Diop, the main street, cutting off traffic flow and access to certain spaces in the city. The latter is symbolized by the 1988–1989 social movement, Set/Setal ("Be Clean/Make Clean"). Here, city residents "set out to clean the city, buttress the failing urban waste infrastructure, and purify a polluted political sphere" (Fredericks 2018, 2). Whether dirtying or cleaning the city, Dakar's urban communities have used garbage as a signifier of their common aspirations for their city. Fredericks argues that in the context of Dakar's "culture where cleanliness of body and soul is of deep spiritual import, their acts of dirtying or ordering public space are profoundly meaningful" (Ibid., 3). This assertion helps illustrate the role of junk in the public space in conveying underlying urban ideas. Although Fredericks used these garbage scenarios to discuss political aspects of urban citizenship, her work motivates another discussion of garbage in cities; that is, how garbage has become an infrastructure of articulating the crisis of the urban commons in Africa. Giving prominence to the garbage and the communities who reside in such spaces, this chapter proposes trash as a valuable symbol of articulating the crisis of urban commons. When applied in the interpretation of *Nairobi Half Life*, garbage indexes the challenges of dejected people keen to traverse between their marginal spaces and the main city.

Like many postcolonial African cities, present-day Nairobi has continued to urbanize rapidly and haphazardly, with greater percentages of the urban population residing in informal settlements (Davis 2006; Herr and Karl 2003, 19). A spontaneous and highly segregated growth pattern of slums and exclusive elite enclaves, archetypal of postcolonial urban Africa, has taken the place of orderliness and improved infrastructure (Caldeira 2016; Miraftab and Kudva 2015, 93; Pieterse 2011, 3). Among other consequences of the city's differentiated form is the difficulty of accessing mainstream benefits, such as markets, amenities, and formal capital systems. The slum, sometimes the only chance for urban belonging, maps out a tacit negotiation that does not guarantee inclusion in the benefits of common urban resources, but nevertheless offers a chance to stake out a portion of Nairobi city, where the struggle for common access and use of the city is far from a petty problem.

Nairobi's urban commons

In May 2020, amid the Covid-19 lockdown in Nairobi city, a residential portion of the neighbourhood of Kariobangi, Nairobi, was demolished to give way to infrastructure for an upcoming affluent project. For days and nights, displaced residents camped outdoors. Several days later, another residential space in Nairobi's Ruai estate was bulldozed at night, again dispossessing residents. In June 2020, parts of Gikomba market were burned down (Nation TV 2020). Within days, when the traders had rebuilt their stalls, government security agents demolished them again. Lamenting this incident, the area member of parliament, Charles Jaguar (2020a), wrote on Twitter:

> NYS recruits are in Gikomba right now. They're demolishing stalls that traders had reconstructed after a mysterious fire had razed them on Thursday morning. GSU officers are also there protecting the recruits. Is this exercise a sign that the govt was involved in the recent fire?
>
> (Jaguar 2020a)

He was subsequently arrested for protesting the demolitions (Jaguar 2020b).

Such demolitions of city spaces – now a fairly recurring practice in Nairobi – index the ongoing competition for the occupation of Nairobi city (Mututa 2019). They also demonstrate the in situ mechanisms that govern how the city can be shared among its different socio-economic classes of residents. "Who can own which space?" and "What can be deemed common versus private spaces in the city?" – questions echoing Léfebvre's (1991) theories of space and rights to the city – are possible perspectives through which to understand these impromptu displacements in Nairobi city. A possible response is to be found in Musambayi Katumanga's (2005, 518) argument that Nairobi is undergoing "constant de(re)composition … characterized by vicious struggles over spaces for socio-economic reproduction." "De(re)composition" here is a binary term which aptly expresses the city's polarity. On one hand, decomposition suggests a city undergoing degradation, in this case led by fraudulent economic networks keen to amass a real estate portfolio through clamour for strategic city spaces. It is characterized by lethargic development, monstrous corruption, and diminishing publicly available spaces. The most visible form of this process is the destruction of existing urban structures to give way to newer modern ones. In short, the trashing of one urban form to pave way for another becomes a paradigm of an ongoing process to convert Nairobi from a common space to private equity. On the other hand, recomposition suggests both the emergence of new urban forms in place of the demolished ones as well as the capacities for resilience within such environments. Thus, whereas demolitions in postcolonial Nairobi have become common occurrences, the consequence is not only outright engineering of new spatial practices but also the emergence of new strategies to cope with such displacements. The capability of residents to mobilize against such displacements, even when futile (as was the case with recent demolitions cited above), signals a desire to fight back and repossess a quota of the city. This crisis cannot be separated from the discussion of Nairobi's postcolonial crisis urbanism.

What is revealed, and what this chapter pursues, is not merely greed for real estate portfolio acquisition. Such demolitions pose the resulting spatial conflicts as a form of crisis urbanism in the context of inhabiting the city as a common space. The displaced residents squatting for days in the ruins of their demolished structures in Kariobangi and Ruai and the informal traders rebuilding their structures in Gikomba despite the presence of government security agents epitomize an entitlement to a part of the city (Chappatte 2015, 9). Accordingly, their pursuit of rights to these spaces is not a knee-jerk manoeuvre but a signal of their desire to inhabit the city as a site of the commons (Borch and Kornberger 2015).

The idea of the urban commons derives from the broader concept of the commons as a depleting resource (Hardin 1968), hence increasingly unavailable for common use (Ostrom 1990). Urban commons refers to the resources commonly available to city inhabitants, such as real estate, that diminish with occupation, just as consumables diminish with utilization. A space occupied by one city resident becomes unavailable to another and is equally subtracted from the total space available to the rest of the city residents. This produces a tense relationship between various urban residents competing for spatial occupation. It is also the normal form of urbanism for many Nairobi residents. By this rationale, then, crisis urbanism resulting from the pursuit of the urban commons refers to the difficult decisions inherent to residing in a city where belonging is constantly tentative. In this scenario, occupying trashed (undesirable) spaces can be construed as a tactic to claim common belonging in a highly fragmented and competitive city. Despite its prevalence as a mainstay aspect of Nairobi's postcolonial urbanism, this critical manoeuvre has not been well addressed. Thus the question pursued here is: how do such happenstances support a theorization of the urban commons in postcolonial Nairobi?

This question intimates that Nairobi's urban commons can be critiqued through, among others, the parameters of collectivity, government, and inclusion or exclusion (Borch and Kornberger 2015, 1). The underlying logic is that urban communities do not gain equal collective usage of city spaces but are subject to various mechanisms that regulate capacities to use the city as a shared resource. To belong to an urban community or to occupy a specific urban space is not a straightforward manoeuvre. It involves underlying mechanisms that designate which streets and neighbourhoods one may occupy at any given time depending on the social community with which one is affiliated. Accordingly, the spatial routines of such urban communities are apertures into prevalent urban conjectures of what is public (commonly available) or private (reserved for certain categories of urban tenants). Hence, although habitation of the city by a diversity of residents generally marks it as a site of commons, restrictions on usage of spaces and the resulting "profound inequalities" (McFarlane and Desai 2016, 146) alert us to the restrictions of thinking of the city as a common space. The puzzle of contemplating the urban commons has a history: if the city is a place where those displaced from the commonality of rural communal life congregate (Thompson 1963), "how, then, could the city also be a site of the commons?" (Huron 2015, 6). That is, what perspective is appropriate to discuss the urban commons within the context of this paradox?

When we see the urban commons as "not just a commons that happens to be located in a city ... (but) a specific way of experiencing collective work, among strangers, to govern non-commodified resources in spaces saturated with people, conflicting uses, and capitalist investment," and its challenge as "to reclaim the commons: to work with strangers to wrench resources from a capitalist environment" (Huron 2015, 15), the urban commons presents a theoretical framework for how urban tenancy is an array of encounters between city residents working against existing economic interests. The resulting zero-sum crisis is characterized by the informal occupation of the city, which, despite providing an option for urban belonging to those with lower incomes, also negates some of their rights to inhabit the city as a common space. The perspective adopted in this chapter is that the urban commons is a "relational phenomenon" comprised of usage and consumption (Borch and Kornberger 2015, 7). It is a relationship between various city residents united by the possibility of their probable disposal without notice. In postcolonial Nairobi of conflicting spatial interests, such as described by Katumanga, the phenomenology largely implicit in Tosh Gitonga's *Nairobi Half Life* – that is, occupying trashed spaces, in this case, downtown Nairobi – is useful in theorizing the urban commons as a phenomenon of crisis urbanism.

Nairobi Half Life

Nairobi Half Life is the story of Mwas (Joseph Wairimu), who resides in Gaza, a dilapidated space in downtown Nairobi. Early in the film's first sequences, he travels from his village to Nairobi seeking an acting career. From his initial characterization as a hawker selling pirated digital video discs in the village township, Mwas typifies the illegality associated with urban struggles and resilience of urban life. His journey to Nairobi is shown in an elaborate scene, starting with his pre-departure moment with his family the previous evening; capturing his journey along the dusty village roads where he bids a neighbour goodbye; and culminating in an extended sequence of the road journey that ends with his arrival and alienation in Nairobi. This emphasis on his journey to Nairobi produces a linear temporality, which then becomes helpful in mapping out the city from his stranger's perspective. The film's idea of how access to the city is controlled is equally given importance, with different powers competing to control the spaces. There are the downtown gangs who accost Mwas soon after he disembarks and rob him of all his goods. Then there are the hawkers whose ephemeral presence in the streets is poised as a disruption of the city's formal order. Subsequently, we see the city's council officers as another urban force that Mwas is made to recognize after an impromptu arrest. Later, we have the police officers who restrict urban access through incarceration of suspects. Finally, there is the downtown criminal gang whose leader, Oti (Olwenya Maina), whom Mwas encounters in jail, and who recruits him into the downtown city, christened Gaza. Oti's gang resides alongside Dingo's (Abubakar Mwenda) gang in Gaza. Gaza's other inhabitants include prostitutes, car vandals, petty thieves, and informal entrepreneurs, including food vendors and vandalized vehicle-parts resellers. In this film, Nairobi city is divided into the formal uptown city, where police officers and other formal businesses reside, and the informal downtown, where the city's poor communities, personified by the gangs and the protagonist, occupy junk vestiges like Gaza in downtown Nairobi. While uptown is framed to accentuate modern streets and high-rise buildings, the film's mise-en-scène of Gaza – the film's dominant setting – is comprised of dilapidated spaces: dirty streets, hotels constructed with corrugated iron sheets, tiny and overcrowded sleeping rooms, the dirty river, messy spare parts vendors, and a junkyard.

This visual depiction of Gaza as a territory delinked from the main city – its geographical location contiguous to the main city yet experientially excluded from its benefits – and the unequal power relations between its residents and those of uptown Nairobi, conjure notions of the crisis of the urban commons in Nairobi. Gaza is, for the characters, a liminal space between a version of the city they want to discard and one to which they aspire but which remains unattainable. It is a site of ruins where those displaced from the city of their desire congregate. Gaza is thus a metaphor of unequal access to Nairobi's allure of economic opportunities for its residents. Nairobi here is not just the physical space comprised of roads, streets, and avenues; residential and commercial buildings; means of mobility; or utilities. Rather, it is primarily the non-material socio-economic map comprised of different urban communities that coalesce in various spaces and imbue them with meaning through daily consumption of various city resources. In this film, Gaza is important as a site of critiquing the urban commons for two reasons.

First is Gaza's narrative valence as a symbol of an omitted community. A recent critique of *Nairobi Half Life* refers to Gaza as a heterotopic space (Mututa 2019), suggesting its inhabitants are cast off from the main city. Gaza's residents thus convey the crisis of Nairobi's postcolonial urbanism through their material precarity and acts of junk salvaging and reselling of vandalized car parts. They also convey the class struggle evidenced by Mwas' ambition for an acting career uptown, and Oti's girlfriend, Amina (Nancy Wanjiku Karanja), who aspires to be an entrepreneur. Further, Gaza typifies the identity of an urban outlaw through illegal entrepreneurship depicted by vandalizing, petty street theft, drug peddling, and prostitution. In these aspects, Gaza epitomizes an epicentre of Nairobi's crisis urbanism.

Second, Gaza offers a reference point for critiquing the urban commons, both through the harsh life endured by its residents and their contact with residents from uptown. As Mwas' character trajectory – from village dreamer to member of the urban poor – adduces, postcolonial Nairobi is a composite of these two disparate parts. Not only that, but also that through him, the film "bring[s] to life, on the screen ... characters who confront the harsh and sometimes deadly realities for those living on the margins of large urban agglomerations in the global south" (Bouwer 2019, 69). There is the main city that remains unattainable to him throughout the film. The closest Mwas comes to this version of Nairobi is in the theatrical play incorporated within the film, where he breaks into an affluent uptown home as part of his stage performance. There is also the downtown city, where he resides with the criminal gangs. The prevalence of gangs as the dominant figures in the film and their incapacity to enter the main city and own any of its parts register their limits in inhabiting the city as a common space. Backstreet drugs and prostitution deals, the lowly eateries, the crowded sleeping rooms, garbage sites, and abandoned warehouses functioning as execution chambers are all fundamental prisms into Gaza's crisis as a problematic zone, one cast off from the concerns of the main city. In these two aspects, the mise-en-scène of Gaza in *Nairobi Half Life* – which is primarily portrayed through waste in various forms (street dirt, rusty eateries, murky river, smoke pollution, scrap metal yards), and the actions and activities symbolic of urban decay (prostitution, crime, corruption, exploitation, murder, drug peddling) – designate trashed spaces as a critical metaphor of the spatial imaginary and representation of ideas around Nairobi's urban commons in this film. This can be inferred from the use of trash as a pervasive symbol of the crisis of the urban commons in Nairobi cinema.

Garbage in Nairobi cinema

Representations of garbage are a common aesthetic in many Nairobi cinema narratives. Muchiri Njenga's *Kichwateli* (2012) tells the story of a young boy, Kichwateli (Carlton Namai), who traverses the city–slum frontier. The film starts with a night scene of the young protagonist framed as a forlorn silhouette walking past a crow perched on a dried tree stump in a barren space, the dark city forming the background. At the start of this scene, the camera pans left as the feet of the young protagonist enter the frame, silhouetted images of city buildings visible on the horizon. The second shot in this scene is a wide composite shot of the protagonist walking in a barren city edge, framed at a low angle. In this shot, the camera zooms out to highlight a subjective view of the protagonist as he walks towards the city's modern skyline visible in the background. Noticeably, the slum from which he comes is omitted from this initial framing of the city.

This shot is emblematic of the way the slum is a space subtracted from the city. The composition of the protagonist in the foreground and the city in the background, and the use of reverse kinesis – whereby the protagonist is moving from the foreground towards the background – index his desire to rise above the limitations of slum life. The dark colours and shadowy lighting used in the shot – both aspects of noir aesthetics, imbue the scene with a sense of apprehension, foregrounding the protagonist's alienation from the city. The image of the crow juxtaposed with the protagonist further heightens his sense of loss and abandonment, persuading the viewer to see the slum, which he personifies, as an inconsequential urban space. The composition of the opening montage showing a protagonist overwhelmed by dark shadows, open spaces, and shadowy lighting conflates two useful metaphors: that the slum community is edging towards the city, and that it is doing so with perceptible apprehension. Further, the image of the crow, a bird that typically resides in city dumpsites, adds to the signification of the absent slum as a dumpsite.

Equally significant is that this opening montage transitions to the protagonist's hasty sojourn through the city, followed by a dream scene in which his head transforms into a television set improvised from assembled junk. Joshua McNamara (2013, 129) acknowledges the junky television head thus: "a young urban adventurer undertakes an odyssey from Kibera to downtown Nairobi, wearing a jumpsuit and a television on his head." He attributes the narrative conflict to Kichwateli showing urban residents the "images they don't want to see" (Ibid., 131). We can thus not discuss Nairobi city in *Kichwateli* without acknowledging the metaphor of urban junk which is central to this film's material symbolism as evidenced by the main character's head which is fashioned from discarded junk (Studio Ang 2012). A further description on the same website page gives character to the junk: "the hanging jack-pins, cables and wires were a subtle metaphor for dread-locked hair representing people of African descent particularly freedom fighters." This description humanizes urban junk and gives it character and mobility. The protagonist as a human character is supplanted with the television head, thus positing junkiness as the primary attribute of the urban community comprised of lower-class urban residents that the protagonist personifies.

From this construal, a reading of Nairobi's urban commons in *Kichwateli* is also possible. In this film, Kichwateli embodies the reconstitution of urban waste into a different form and utility. Considering that the television set displays information about happenstances or news, we can read the protagonist's head, his sojourn across the city, and his point of view of the city as a critical perspective on the everyday life of those existing as "trash" in Nairobi. His inability to fit in the city provides a useful template of Nairobi's polarity, subsequently visualized by a wide composite shot of Nairobi's slums and the city centre.

This shot is comprised of a foreground of rusty roofs of slum structures and a background showing Nairobi's modern skyline of high-rise buildings. The middle ground appears indistinct. This composite shot of Nairobi augments the mise-en-scène of distinctly different material landscapes of the city to suggest Nairobi not as an "aesthetic whole" (Whiteley 2011, 41–42) but as a composite of fragmented spaces and communities. Further, such composition supports an argument that *Kichwateli* is about Nairobi's split into a motley city and not its unification. The shot's material infrastructure highlighting the slum can thus be said to sentimentalize it as a misfit amidst the city's glamour and not merely positioning it as a periphery. The rusty slum here appears as a "'forefront' – a starting point" (Rahamimoff 2005, 69) for a version of Nairobi detached from the modern city. Its material decadence identifies it as the city's informal appendage, appropriately designating it as an urban existence that is both cast away from the city's modernity and a space upon which the crisis of Nairobi's urban commons is most evident.

This argument about the crisis is further deducible from the cinematography of the shot which uses a sepia colour tone – marking the composite city as barren and extraneous. The sepia tone suggests the idea of an amalgamated Nairobi, comprised of the slum and the modern city, as a fading, decaying one. Furthermore, the use of a convergent lens to create a distorting perspective, whereby the modern city appears unstable as the buildings appear to tilt, suggests the instability of such a divided city. Thus, the high-to-low composite camera angle highlights used in the shot designate the city's sharp divisions into unequal segments. The foreground slum, framed from a high angle, appears subdued yet hypervisible. The background city, framed from a low angle, appears dominant, yet its horizontal position as the background designates it as less prominent. Such a shot clearly expresses the paradox of Nairobi's urban commons: that a city foremost experienced as a slum by a majority of its residents is now tasked to signify the slum's exclusion from the commonality of urban life.

The question then is: what does this composition of Nairobi and the mise-en-scène of the rusty slum convey about relations of dominance between the foreground slum and the background city? The shot's alternating high and low camera angles constitute a simultaneity of a beleaguered and inconsequential yet visually dominant slum and a materially superior city lingering in its background as a visual metaphor for how Nairobi continues to be experienced as a binary city. In other words, the main city and the slum are contiguous yet restricted. It further detaches the symbolism of the urban commons from framing and composition (which are certainly contradictory) to the visuals of trash or its absence. If superimposed with the film's opening shot, the foregrounded slum would correspond to the barren land visually assigned to the protagonist, while the city would retain the background position. It thus follows that the framing technique used here posits the slum's rusted material structures as metonymic of poor urban residents, such as Kichwateli, who have no real space in the city, and the skyline of modern buildings as metonymic of the modern city. Certainly, through this juxtaposition of the slum – a trope of urban junky spaces – and the city – a symbol of modernity – the film caricatures Nairobi's incompatible outline (Léfebvre 1991), whereby the two spaces coexist in a continuum but do not allow common access and usage. The underlying urban condition – clearly differentiated by the material infrastructure of dilapidation versus opulence – marks the protagonist's crisis of traversing between the two spaces.

Whatever other conversations such mise-en-scène and composition in the scenes discussed above may enable, the protagonist's characterization as garbage maps out the narrative value of trash as a "lens into questions of urban citizenship" (Fredericks 2018, 3). Kichwateli's fleeting journey through the city and his restricted access to the main city in the final sequence of the film augment the film's visual composition to render symbolic experiences within Nairobi city and raises important questions on urban survival. It is noteworthy that Kichwateli's walk across Nairobi traces the difficult terrain of urban junk – and the poor urban community that it indexes – across a spectrum of space and time: the slum, the streets, and the city. He elicits curiosity and a sense of awe in the slums and rejection within the modern city. The film ends with Kichwateli being chased away from the formal city by a security officer and a mob, speaking of the elaborate systems restricting access to certain spaces in the city for those designated as urban junk.

Diego Quemada-Diez's *I Want to be a Pilot* (2006) uses more explicit images of garbage. It juxtaposes a young male protagonist, Omondi (Collins Otieno), with a dumpsite located in the slums. The first notable feature of this film is that it is narrated in the first person, marking the story as the protagonist's personal narrative of the city. The film's opening scene shows the protagonist and other pupils inside a dingy classroom located within the slums, then a shot of an aeroplane flying by. The subsequent shot within this montage is a high-angle close-up of the young protagonist looking at the aeroplane. By establishing the film with a shot of a protagonist living in the garbage heaps and contrasting it with the aeroplane, this montage forms a baseline upon which critical urban ideas may be deciphered. First, it communicates the protagonist as urban poor. His residence in the garbage dumpsite and the poor condition of the school where he learns are also posited as urban realities of his immediate space. Second, his juxtaposition with the aeroplane frames both his aspirations for a better life and affirms his limitations to achieving this ideal. That he keeps wishing to be a pilot – a dream that seems, within his squalid space, very remote – indexes the imbalanced opportunities between the city's poor and affluent residents. Thus, whereas contrasting the modern city inhabited by the pilot with the squalid slum inhabited by the protagonist clearly separates the city into unequal fragments, this binary also underscores the pressing issue of how occupying certain parts of the city – in this case, the slum – confers upon certain residents restrictions in the city. Here, the protagonist's inability to access the main city (and its opportunities) is conveyed through the metaphor of garbage.

In Nathan Collett's *Kibera Kid* (2006), a garbage site in Kianda, part of Nairobi's Kibera slums, is the main setting of the story about a young boy, Otieno (Ignatius Juma). The opening sequence starts with a wide-angle panoramic long shot of Nairobi's modern skyline. The camera then swish pans to a high-angle shot of rusted iron sheet roofs over mud-walled shacks in the Kibera slum. The film uses the screen titles "Nairobi, Kenya" for the city shot, and "2 Miles South, Kibera" for the shot of Kibera slum. While the swish pan creates a sense of proximity between the two spaces, the subtitles "Nairobi" and "Kibera," respectively, metonymically underscore the disconnection between the two spaces curated by the mise-en-scène of the two adjacent shots, comprised of modern high-rise buildings and the rusty material landscape of the slum and an adjacent garbage heap. The viewer thus watches the protagonist's urban squalor and his final journey from the slum towards the city with an awareness of the initial glimpse of a progressive city as a space whose opportunities are not commonly available to slum residents. Although Otieno constantly shows his talent and dream of becoming a musician, he has no chance of actualizing this dream in the slum. Instead, Otieno's character arc develops from scavenging in the dumpsite to being recruited by Razor, a criminal gang operating in the slum. He becomes the front runner of a vicious war between the gang and the community, led by Wamatope. The turning point is when he fails to kill Wamatope, despite orders from his gang boss, J L (Godfrey Ojiambo). Cornered by an angry lynch mob, Otieno is rescued by his would-be victim, Wamatope, who advises him to leave the slum to pursue his music dream in the city. With this narrative plot point, the swish-pan transition between the two initial shots – mimicking a swift turning of the head away from the modern buildings of the city to the mud-walled rusty tin-roofed shacks of Kibera slum – can be said to draw attention to the significance of the "two miles" separating the city from the slum. This archetypal characterization of Otieno's slum life draws attention to the slum as an extension of the garbage heap. The garbage heap designates not just the physical putrefaction of the slum space but also its limitations in terms of accessing the city and its opportunities.

At a time when slums are increasingly the primary form of urban expansion in Africa (Davis 2006), these three short films provide a baseline for discussing images of garbage as metaphors of the urban commons in Nairobi. On one hand, we can read the images of garbage (and hence slums and slum communities) as exemplifying the "material processes of abjection through which certain bodies become constituted as waste" (Fredericks 2018, 17). Here, mechanisms of classifying and excluding certain groups of urban residents from wholly and transparently participating in or claiming ownership of urban opportunities and spaces simultaneously construe a crisis of accessing the city as a common space. On the other hand, highlighting images of squalor reveals the "economic, historic, social, and political forces that are largely unseen to the casual observer" (Gilderbloom 2008, 1). The implication is that the mostly exclusive access and usage of Nairobi city envisaged in these films historicize crucial urban structures. Appropriately, these films lay the groundwork to read images of garbage in *Nairobi Half Life* as usefully representative of postcolonial crises of the urban commons in Nairobi.

Urban commons in *Nairobi Half Life*

In *Nairobi Half Life*, images of garbage appear early in the film. One of the most memorable scenes occurs when Mwas is incarcerated overnight in a city jail. Here, he is bullied into mopping the common toilet space whose floor is full of faecal filth. During this process, he falls on the filthy floor and contaminates himself with the faeces and urine. Curiously, he is unbothered and starts singing as if celebrating his filthiness. In this scene, faeces acquire the agency of Mwas' transformation into an embodiment of urban filth. Thus, when he is released from jail and walks to downtown looking for Dingo's gang, we can easily connect his personification of filth with downtown as his assigned urban space. From the deixis of the subsequent mise-en-scène of downtown, a paradigm of urban trash emerges.

The first time Mwas meets the gang they are inside a downtown shack. The mise-en-scène of this scene is comprised of salvaged pieces of junk: repainted iron sheets form one of the shack's two walls, an old polythene paper and crooked wooden poles are partially visible in the foreground, and an old charcoal stove lies at the edge of the shabby sofa. Subsequent medium shots show a haphazardly arranged repository of salvaged wrecks: a worn-out car door, mismatching pieces of the grill, a worn wooden sofa and a cement block, boxes piled on one edge of the room, motor vehicle reflectors randomly fixed on the iron sheet walls, an old power supply unit nearby, an old wooden seat, and what looks like a support column for an once-executive office chair. Such mise-en-scène characterizes downtown as a place crafted from the city's huge leftover market, what Abraham Akkerman and Ariela Cornfeld (2010, 34) call "street hardware." The significance of this composition of Dingo's downtown shack is found in the juxtaposition of reassembled garbage and downtown characters, conferring upon the characters the same attributes as the garbage, so that we see them as part of the trash structure. Through this association, trash acquires the metonymic value of Nairobi's downtown community, and its condition of decay indexes their urban condition. Gaza, the downtown space where the structures and the characters exist, inherits the attributes of undesirability, exclusion, and expendability.

This characterization of Gaza in *Nairobi Half Life,* using metaphors of urban rejects, indexes everyday urban crisis; namely, the inability of downtown residents to benefit from the city. Mwas' struggles in the city, evidenced by his crime and difficulties in securing an acting career in an uptown theatre, underscore the difficult decisions he has to make if he is to reside in the city. Furthermore, the initial events leading to this point – a daylight attack by a criminal gang, an arrest by the city council security officers, and unlawful overnight detention in a police station – all serve to alert him to the fact that the city is not a common space, but one restricted and regulated by different groups. In the character journey of Mwas, then, various crises of Nairobi's urban commons become apparent. These include economic exclusion, spatial restrictions between downtown and uptown, and power hierarchies between uptown police and downtown gangs.

Further, this composition of garbage dramatizes everyday urban life in Nairobi's downtown, forcing the viewer to consider the uneasy relationship between city inhabitants as exemplified by the tension between uptown police and downtown gangs. That the downtown junkyard precedes its human characters, who only encounter it and use it to construct their spaces, and that it precedes Mwas who, despite being the protagonist, is inserted into a junky space, signals the dominance of garbage over those who inhabit it. It both characterizes and determines the portions of the city that are accessible and usable by those who inhabit downtown Gaza in the film. Garbage here has a performative role; it elicits the ominous conditions of this part of the city and challenges Mwas' earlier notion that Nairobi is a place of commonly accessible opportunities. "What does this do?", we may ask. In contravening such expectations – that anyone can freely achieve their dreams in the city – the junkyard pre-empts other pressing questions, such as where can Mwas thrive? If all Nairobi is not accessible to him, what marks the default spaces where he can fit from the others where he must not? There are three possible arguments that follow.

First, that *Nairobi Half Life*'s use of the aesthetic of garbage to signify the difficulties of sharing the city suggests that Nairobi's downtown is an integral but different part of the city (Roy 2011). It is a common ground insofar as those with minimal economic capacities can stake a claim to the city but a marginalized zone in that it does not afford its tenants appropriate opportunities for self-actualization. The gangs residing here vandalize cars uptown and sell the parts downtown. Prostitutes here receive income from uptown clients. Yet, these activities are not freely practised, as corrupt police officers from uptown visit downtown to receive bribes from the gangs. Through these relationships, the film creates a one-way border, whereby downtown is neither too removed from the main city, nor included in its benefits, but instead appears as a city in waiting: "one that constantly lives under specific threats and incompletion" (Simone 2016, 136). To inhabit downtown is to inhabit a liminal space, to exist with uncertainty over what could happen from one moment to the other. Downtown's images of junk are thus multipurpose registers suggesting that downtown is not an uninhabitable space (Simone 2016), or a place for the "marginal and insignificant" (West-Pavlov 2005, 43), but the real face of the city from which it is excluded (Pieterse 2011, 6). Here, those unable to afford urban residence elsewhere "gain a foothold as normative citizens" (Simone 2016, 139) of the city. Accordingly, junk connotes a struggle to define the limit of the commons in Nairobi city. While downtown cannot be wholly subtracted from the rest of the city, the main city cannot exist as an autonomous entity from its downtown communities.

Second, from its signification of spatial marginality, we can see downtown as an inventory of the tensions of urban coexistence and the manoeuvres that become necessary to inhabit such a historically polarized colonial city as Nairobi. Nairobi's centre-edge design and the city's mixed political economies have, since the colonial period, created grounds for sustaining spatial displacements. The most visible symbol of this trend is Kibera slum. Founded for Nubian ex-soldiers after the World War II (Desgroppes and Taupin 2011), Kibera has grown to become one of the biggest informal urban settlements in Africa. Yet, curiously, until Mikel Maron and Erica Hagen launched a community-based mapping project of Kibera slum in November 2009, it remained largely a blank spot in the global maps (Haklay et al. 2014, 44). That these scholars used Kibera's geographical invisibility in Nairobi's digital geographical map of the time to suggest its signification of urban marginality provides a useful framework to study the way material attributes of specific urban sites may signal the underlying crisis of urban belonging. Thus, beyond the obvious sidelining of the slum from the city's gentrification policies, the slum may be seen as a placeholder for a continuous negotiation around urban residency. It is a material signifier of the existential conditions of the urban poor and hence defines not only a condition of dejection but also implicates the beneficiaries of such condition (Harrow 2013, 1).

The junky downtown in *Nairobi Half Life* construes a space where those excluded from the main city accumulate "out of sight" yet retain contact with the city. It is a holding zone, a sort of frontier that identifies the limits of accessing the main city or claiming inclusion into its opportunities and benefits. This film's visual aesthetics of Mwas' initial entry into the main city illustrate this point. In the first two medium shots after losing his property to robbery, Mwas' framing alters the depth of focus, isolating him from the city in which he walks. Defocusing the space around him detaches him from the city space around him, marking him as different from the urban community he is walking within and the city space he has accessed. Further, this blurred view of his surroundings suggests his disillusionment, identifying him as a misfit. Even when the framing shifts to wide shots and deep focus shots of Mwas as he makes his way further into the city, the film uses high-angle framing and continuous zoom-out camera movements to show his insignificance. These slow zoom-out camera movements are stylistic choices that show Mwas as an instrument of embarrassment from which the camera retreats. The change of this initial visual style when Mwas arrives in Dingo's shack marks downtown as his allotted territory within the city. The idea of the urban commons pursued here through the junky shack is that which speaks to the scarcity of "inclusive urban commons, or accessible urban spaces that facilitate a hybrid mixture of social exchanges between individuals and different social groups" (Murray 2011, 215). When we see the old sofas being reused as seats, the old iron sheets as new walls, and the crooked poles as support columns of the shack, we see such reusage as attempts to enable new forms of coexistence within the city.

The third argument is that the characterization of downtown residents using metaphors of garbage and the opportunistic manner in which these characters chance upon their livelihoods within the city retrieve the imagery of dispossession, poverty, social immobility, chaos, crime, and other social ills as inalienable forces that shape urban conditions in postcolonial Nairobi. The arising matter here is no longer urban survival but rather the agency of such a space in pre-empting a strand of crisis urbanism. By existing as a structure that contravenes the expected form of urban tenancy, Dingo's shack designates a material inconsistency, a gap, between downtown and uptown. This space is not a "neutral void in which objects are placed and events happen" but a "medium with its own consistency and, above all, its own productive agency" (West-Pavlov 2009, 17). Its form as an aggregate of junk gives agency to downtown residents to convey their tacit struggles to cross the "gap" from one version of Nairobi to another. Instead of merely characterizing the constant struggle of downtown, Dingo's shack "stresses the presence of margins without revealing its contours" (Chappatte 2015, 7). It traces the urban commotion at the frontier of the informal and the formal city where the residents are constantly struggling to commonly share the city.

The three arguments raised above, all focusing on the agency of garbage in its raw or symbolic form, mark the "protracted struggle over the … right to survive in the city" (Simone 2004, 169). We see, in Mwas' circulation, an attempt to embed and flourish in a city from which he is excluded. Crimes such as vandalism, prostitution, drug dealing, and car theft are not merely avenues of "primitive accumulation" (Katumanga 2005, 515) but a "vehicle through which … collaboration emerges" (Rahamimoff 2005, 81) between the city and its margin. The ubiquitous interactions between downtown gangs and uptown police and the downtown car vandals and the uptown car owners are in situ collaborations that pelt legal against illegal, lawful against lawlessness, survival against death, morality against narcissistic corruption. Through such conflicts, we see Nairobi city as a "gigantic assemblage of junk continually being re-made and re-inscribed" (Whiteley 2011). It is where Nairobi's crisis of sharing the city in common produces what Orvar Löfgren (2015, 68) terms as communalities, "where people with very different backgrounds mingle" and "different users co-inhabit and regulate public space." Communality here refers to the downtown and uptown characters who encounter each other in the downtown space, which is thus where the crisis of the urban commons in postcolonial Nairobi is most recognizable through the resulting communality.

Bibliography

Akkerman, Abraham, and Ariela F. Cornfeld. 2010. "Greening as an Urban Design Metaphor: Looking for The City's Soul in Leftover Spaces." *Structurist* 49/50: 30–35.

Borch, Christian, and Martin Kornberger. 2015. *Urban Commons: Rethinking the City*. Oxon and New York: Routledge.

Bouwer, Karen. 2019. "Life in Cinematic Urban Africa: Inertia, Suspension, Flow." In *A Companion to African Cinema*, edited by Kenneth W. Harrow and Carmela Garritano, 69–88. Hoboken: John Wiley & Sons.

Caldeira, Teresa P.R. 2016. "Peripheral Urbanization: Autoconstruction, Transversal Logics, and Politics in Cities of the Global South." *Environment and Planning D: Society and Space* 35 (1): 3–20.

Chappatte, André. 2015. "Unpacking the Concept of Urban Marginality." *ZMO Programmatic Texts* (No. 10) Zentrum Moderner Orient. https://www.zmo.de/publikationen/ProgramaticTexts/chappatte_2015.pdf.

Davis, Mike. 2006. *Planet of Slums*. London and New York: Verso.

Desgroppes, Amélie, and Sophie Taupin. 2011. "Kibera: The Biggest Slum in Africa?" *Les Cahiers de l'Afrique de l'Est* 44: 23–34.

Fredericks, Rosalind. 2018. *Garbage Citizenship: Vital Infrastructures of Labor in Dakar, Senegal*. Durham and London: Duke University Press.

Gilderbloom, John Ingram. 2008. *Invisible City: Poverty, Housing, and New Urbanism*. Austin: University of Texas Press.

Haklay, Mordechai, Vyron Antoniou, Sofia Basiouka, Robert Soden, and Peter Mooney. 2014. *Crowdsourced Geographic Information Use in Government*. Washington, DC: International Bank for Reconstruction and Development/World Bank.

Hardin, Garrett. 1968. "The Tragedy of the Commons." *Science* 162 (3859): 1243–1248.

Harrow, Kenneth W. 2013. *Trash: African Cinema from Below*. Bloomington and Indianapolis: Indiana University Press.

Herr, Harvey, and Guenter Karl. 2003. *Estimating Global Slum Dwellers: Monitoring the Millenium Development Goal 7, Target 11*. Nairobi: UN-HABITAT working paper.

Huron, Amanda. 2015. "Working with Strangers in Saturated Space: Reclaiming and Maintaining the Urban Commons." *Antipode* 47 (4): 1–17.

Jaguar, Charles. 2020a. Twitter Post. June 27, 2020, 12:38 a.m. https://twitter.com/ RealJaguarKenya.

Jaguar, Charles. 2020b. Twitter Post. June 29, 2020, 12.34 p.m. https://twitter.com/ RealJaguarKenya.

Katumanga, Musambayi. 2005. "A City Under Siege: Banditry & Modes of Accumulation in Nairobi, 1991–2004." *Review of African Political Economy* 32 (106): 505–520.

Léfebvre, Henri. 1991. *The Production of Space*. Translated by Donald Nicholson-Smith. Oxford: Blackwell.

Löfgren, Orvar. 2015. "Sharing an Atmosphere: Spaces in Urban Commons." In *Urban Commons: Rethinking the City*, edited by Christian Borch and Martin Kornberger, 68–91. Oxon and New York: Routledge.

McFarlane, Colin, and Renu Desai. 2016. "The Urban Metabolic Commons: Rights, Civil Society, and Subaltern Struggle." In *Releasing the Commons: Rethinking the Futures of the Commons*, edited by Ash Amin and Philip Howell, 145–160. London and New York: Routledge.

McNamara, Joshua. 2013. "Thoughts on a Curation of 'The Political' in Film: The 'Filming Tomorrow' Seminar at Film Africa 2012." *Journal of African Cultural Studies* 25 (1): 128–132.

Miraftab, Faranak, and Neema Kudva. 2015. *Cities of the Global South Reader*. London and New York: Routledge.

Murray, Martin J. 2011. *City of Extremes: The Spatial Politics of Johannesburg*. Durham and London: Duke University Press.

Mututa, Addamms Songe. 2019. "Vertical Heterotopias: Territories and Power Hierarchies in Tosh Gitonga's Nairobi Half Life." In *Narratives of Place in Literature and Film*, edited by Steven Allen and Kirsten Møllegaard, 158–170. New York and London: Routledge.

Nation TV. 2020. Twitter Post, June 27, https://twitter.com/ntvkenya/status/12769449 75239004160?s=19.

Neuwirth, Robert. 2005. *Shadow Cities: A Billion Squatters, a New Urban World*. London and New York: Routledge.

Ostrom, Elinor. 1990. *Governing the Commons: The Evolution of Institutions for Collective Action*. Cambridge: Cambridge University Press.

Pieterse, Edgar. 2011. "Grasping the Unknowable: Coming to Grips with African Urbanisms." *Social Dynamics: A Journal of African Studies* 37 (1): 5–23.

Rahamimoff, Arie. 2005. "Jerusalem: Lessons from a Shared City." In *City Edge: Case Studies In Contemporary Urbanism*, edited by Esther Charlesworth, 68–83. Oxford: Elsevier Ltd.

Roy, Ananya. 2011. "Slumdog Cities: Rethinking Subaltern Urbanism." *International Journal of Urban and Regional Research* 35 (2): 223–238.

Simone, AbdouMaliq. 2004. *For the City Yet to Come: Changing African Life in Four Cities*. Durham and London: Duke University Press.

Simone, AbdouMaliq. 2016. "The Uninhabitable?: In Between Collapsed Yet Still Rigid Distinctions." *Cultural Politics* 12 (2): 135–154.

Stam, Robert. 2003. "Beyond Third Cinema: The Aesthetics of Hybridity." In *Rethinking Third Cinema*, edited by Anthony R. Guneratne and Wimal Dissanayake, 31–48. New York and London: Routledge.

Studio Ang. 2012. *STUDIO ANG: Kichwateli*, August 20, https://studioang.com/portfolio/studio-ang-kichwateli.

Thompson, Edward Palmer. 1963. *The Making of the English Working Class*. New York: Pantheon.

West-Pavlov, Russell. 2005. *Transcultural Graffiti: Diasporic Writing and the Teaching of Literary Studies*. Amsterdam - New York: Rodopi.

West-Pavlov, Russell. 2009. *Space in Theory: Kristeva, Foucault, Deleuze*. Amsterdam - New York: Rodopi.

Whiteley, Gillian. 2011. *Junk Art and The Politics of Trash*. London: I. B. Tauris & Co Ltd.

Filmography

I Want to be a Pilot. Dir. Quemada-Diez, Mexico, 2006.
Kibera Kid. Dir. Nathan Collett, USA, 2006.
Kichwateli. Dir. Muchiri Njenga, Kenya, 2012.
Nairobi Half Life. Dir. Tosh Gitonga, Kenya, 2012.

5 Rarray urbanism

The superficies of Monrovia's hustlers in postwar urban crisis

Urban hustles in Africa's civil war cinema

Urbanism in Monrovia, like in many other parts of Africa, is characterized by persistent "postcolonial" crisis. The reference to postcolonialism is, of course, provocative when referring to Liberia, a country where no real colonization happened. In the absence of colonial history, Monrovia presents a unique case of operationalizing crisis as a way of urban life. Monrovia's most notable crisis involved violent seizure by warring factions during years of civil war. In retrospect, the imprint of this history is strongly present not just in the imaginaries of the wartime city but also in the everyday life in postwar Monrovia. Various films set in Monrovia show the city as a watershed for a residual sense of disorientation and displacement, both in physical terms (which was a significant aspect of the wartime period) and also in social, economic, and political terms. This is most evident in the characterization of disenfranchised urban residents who, despite occupying physical spaces in the city, exist superfluously with little connection to its social and economic flows.

This chapter discusses the representation of urban displacement and its operative purpose in the character of the *rarray* – a postwar urban character displaced from social, economic, and political affiliation – as an authentic urban construct. Through a close reading of the rarray in Andrew Niccol's *Lord of War* (2005) and Jean-Stephane Sauvaire's *Johnny Mad Dog* (2008), the chapter explores the rarray as emblematic of west Africa's postwar paradigm of emergent urbanism. The chapter further discusses two key characterizations of rarray urbanism specific to Monrovia cinema: first, in the context of a wartime necropolis and second, in the context of postwar urban fractures. The key argument is that the rarray character is paradigmatic of existential crisis in the city and hence offers an aperture into the superficies of postwar crisis urbanism in Monrovia.

As a region, west and central Africa has had numerous outbreaks of civil wars and armed conflicts. In Nigeria, these span from the Biafra war to the recent Boko Haram insurgency. Sierrra Leone, Liberia, Guinea Bissau, and Côte d'Ivoire have had protracted civil wars in the past. Similarly, the Democratic Republic of Congo (DRC) has seen numerous

DOI: 10.4324/9781003122098-5

armed conflicts. The region's conflicts have caught the imagination of, among others, literary writers and filmmakers. Garrett Batty's *Freetown* (2015), for instance, exemplifies the influence of these wars on literature and cinema. As a background to the discussions of the rarray character in postwar west African urban films, and the rarray's dominance within narrative templates of postwar crisis urbanism, this subsection explores the general characterization of the rarray in Cary Joji Fukunaga's *Beasts of No Nation* (2015), a film adaptation of Uzodinma Iweala's 2005 novel by the same title. Fukunaga's film tells the story of child soldiers fighting with a rebel group to overthrow the government of an unnamed West African country. The narrative plot unfolds around the character of Agu (Abraham Attah), a young boy who flees into the forest when government forces ambush his hometown, killing his family. He is subsequently abducted by a rebel group led by Commandant (Idris Elba) and joins other child soldier recruits fighting their way into the city. Through the overarching motifs of betrayal, greed, and political dismissal that frame the life of fighters, *Beasts of No Nation* positions their disposability in postwar Monrovia both as a way of caricaturing the meaninglessness of war and also expressing its fruitless aftermath.

The film dwells in length on political rifts and government oppression, actuated when government soldiers arbitrarily attack civilians whom they accuse of aiding rebels. By positioning this scene early in the narrative, the film imbues this particular conflict with urgency. One easily notices the precarious nature of civilian life in the city where both rebel factions and government forces pose instantaneous threat to everyday life. The trope of greed for political power is personified by the character of Supreme Commander Dada Goodblood (Jude Akuwudike), who disowns and neglects his fighters once in power, and intimates urban precarity, injustice, and irresponsible political leadership. Further, this character handily exemplifies the proliferation of crisis akin to that which characterized west African wartime cities such as Sierra Leone's Freetown and Liberia's Monrovia. The subsequent retreat of fighters into the forest after being disowned by their political leader, which marks the film's thankless resolution, emblematizes the political expediency surrounding civil wars in urban west Africa and the widespread "social construction of difference between various groups" (Bøås and Utas 2014, 50). It also expresses the postwar condition, which did not transform into sustainable peace (Harris 1999), but rather enabled excessive precarity. This reading builds on some of the recent debates on this film, such as Hilarious Ambe's (2004) review, which focuses on the shot and stench as dramatic strategies that enhance its narration. Particularly, the value of expediency as a postwar urban reality and the kinds of urban practices that it produces in the aftermath of war offer promising starting points to engage with *Beasts of No Nation* as exemplary of west and central Africa's war cinema. To this end, this chapter looks at the way the film constructs the rarray as a postwar urban identity, and through that character, explores postwar urbanism in Monrovia.

Melancholy of dismissal

One of the key moments in which critical ideas of postwar urban identity can be discerned in Beasts of No Nation occurs in the montage of the final sequence of the film when Agu's rebel faction scatters back to the bush, pursued by mortar fire after their political leader has renounced them. From that moment of crisis, the film features a final montage of the rebel group's unceremonious disbanding. This scene is set in the trenches of a non-productive mining site located in the forest. This scene is remarkable because it concludes the narrative of war by framing the unceremonious way in which fighters disbanded. Indeed, when we see these fighters at the war's end – in varying dispositions of discontent and suffering, subsequently arrested by government forces and contained in psychiatric centres – we are reminded of their unrewarded war effort, their lack of formal economic activity, or social community to fall back on, and thus arrives the realization that political betrayal has been the ultimate reward for many fighters at the end of civil wars in the region. The cinematography of the continuous shot starting with Agu walking from a lone sentry firing position, then descending into a labyrinth of trenches filled with brown muddy water, to the moment he exits the trenches at the other end, where fellow ex-fighters are busy "mining" gold, demonstrates useful aspects of war that may guide our understanding of the way the film's elements are working at the level of subtext to express, first, the melancholy of end-of-war dismissal and second, the postwar hustle.

In this montage, the camera is positioned to mimic Agu's eye-view, combined with the shot's composition to convey the elaborate despair endured by the ex-fighters at the end of war. Figuratively, this normal-angle framing communicates the verisimilitude of postwar melancholy. The start of the scene shows a lone sentry fighter firing bullets into an open field. In the absence of an enemy, his action suggests a nostalgia for a war that has come to an end inconclusively. His firing, seemingly without motivation or target, also conveys deprivation when he asks Agu for more bullets, which they no longer have, thus hinting at material precarity when wartime supplies are discontinued. The lone sentry thus personifies the disillusionment of postwar life without economic benefits or political belonging.

As Agu descends into the trenches on his way to the Commander, the camera's eye-level position gives prominence to the brown earthen walls of the trenches, which tower above him, and the knee-deep, brown muddy water, before gradually revealing the rest of his fellow ex-fighters inside the trench, quiescent in various positions of physical exhaustion and despair. We first see a character with a gun hanging loosely under his arm, a prop evocative of his days as an active fighter. As he leans on the trench wall holding his stomach, we are alerted to his famishment and hence the anguish of his postwar life. He has no shirt, and the dirt collecting on his torso and trousers is visible. Juxtaposed with the enclosing earthen walls of the trench, he is seen as a character trapped inside a grave. As Agu proceeds through the trenches, knee-deep in muddy water, we see the rest of his comrades: a delusional ex-fighter smoking while seated in excavated clefts filled with muddy water, unbothered that he is soaked in mud; other ex-fighters sitting chest-deep in the muddy water in complete resignation; and one soldier convulsing, perhaps from hunger, disease, or psychological trauma.

This inventory of ex-fighters suffering in the forest at the end of the war illustrates the collective solitude endured by many ex-fighters at the end of civil war in many regions of Africa. In place of a glorious life in the city, here, the maze of trenches in which they are now immersed is an allegory of the city streets from which they have fled fearing for their lives. This array of disoriented, disillusioned, and disenfranchised characters can thus be interpreted in the context of Monrovia's war aftermath where "remarginalization and not reintegration, ... become(s) the natural outcome awaiting most ex-combatants" (Utas 2003, 250). The images of fatigued ex-fighters reify the verisimilitude of being "stuck, unable to go anywhere, make any changes ... buried under a pile of obligations, mourning, depleted confidence or just too many expectations" (Simone 2011, 43). They have no more political goodwill to continue fighting, they lack material resources to continue fighting or to start a new life, and even their mining venture has flopped. Their despair is also a form of mourning for the meaninglessness of their violent journey into the city, as well as the frustration of their expectations for a glorious reception by their political leaders. As second I-C (Kurt Egyiawan) tells Agu as he dies from a bullet wound, "It was all for nothing." Here he is referring to the war in general and their fighting in support of their political leader in particular.

Postwar hustle

Beyond the allusions to dismissal, the idea of goldmining so explicit in the film's ending deconstructs the meaning of war in the context of hustle. West African civil wars, particularly those located within Liberia and Sierra Leone, have been theorized through the perspective of the fighters as war machinery (Hoffman 2011). However, recent ethnographic studies have suggested that, at least in parts, the Liberian war "was understood primarily as extraction and 'hustle', where extraction displaced violence as the main ingredient of war" (Käihkö 2018, 486). This interpretation benefits the critique of goldmining that has replaced fighting in *Beasts of No Nation's* resolution sequence. In this respect, the scene invites two interpretations.

First, that using the imagery of the trench as a metonymic space of both hope and death, Fukunaga's film paints the picture of an elusive inclusivity. Indeed, he illustrates how the war does not elevate the fighters, but rather buries them in perpetual "suffering and sacrifice" (Waugh 2011, 1), irrespective of whether they continue as fighters or revert to civilian life. The concept of betrayal that surrounds this scene suggests not only war as a process without benefit but also its end as a return to nothingness, as fighters lose various benefits associated with war (Podder 2011, 57). The image of young soldiers wishing to continue the war, some resigned to their fate and others toiling in an unproductive goldmine, constitutes a repertoire of postwar discourses of economic, social, and political marginality. Here, we see the maze of trenches as a figurative mass grave for the ex-fighters. Agu terms his rebel faction as "animals with no place to go" – no war to fight and no political goodwill to incorporate them into mainstream politics. Christensen and Utas (2008, 531) refer to such existence as "animal life," trapped at the interstice of ended combat and unattainable civilian life.

Second, the scene likens the transition from war to postwar as a process whereby one transforms one's labour from merchandising violence in the fields of war, to unarmed forms of hustling. Hustle here suggests the "opportunistic opposite of having a job and steady income" (Käihkö 2018, 496). While the diminished operational capacity of Agu's rebel faction would signal the impracticality of continuing the war, economic liberation – represented by the makeshift non-productive goldmining venture – presupposes (elusive) economic prospects. The trench and the gold venture are evocative of what Christensen and Utas (2008, 522) call the struggle "to establish livelihoods and to manoeuvre within a strictly limited range of peacetime socio-economic possibilities." Limitation in this film partly arises from betrayal and individual aggrandizement and partly from the perception that ex-fighters were essentially dispensable animals. In these shots, the fighters "discarded" inside the trench signify the lost hustle of war and hence their identity as heroes of the past war and the worthlessness of their postwar hustle. The aura of internment, which the film cultivates with the allusion of a grave, is to some extent an inference of the vanity of their conquest. At the same time, their re-marginalization is seen as "justifications and rationales" (Reno 2007, 229) for the possibility of continuing the war hustle, if only to regain the possibility of accessing political and economic spoils. One consequence of this interpretation is that if we see the cities as their barracks (Hoffman 2007) and the ex-fighters as perpetual hustlers in these cities, then "hustling … forms a central idea of what war is to the point of replacing violence as its main activity" (Käihkö 2018, 498). It is in this context, then, that this chapter frames the discussions about this astray ex-fighter character – who inhabits the city as a hustler, the rarray – as indexical of an emergent postwar urbanism in Monrovia.

Rarray urbanism

Christensen and Utas (2008, 517) use the term "rarray boys" to refer to a brand of urban castoffs who thrive precariously outside formal political, social, and economic circulation. In other words, they pass through the city, temporarily establish themselves as forceful custodians of its streets through acts of war, but promptly relinquish this power as soon as the war ends. Rarray is a form of urban citizenship characterized by disorientation and instability, even in the absence of war. First used to describe Freetown's unemployed young toughs who survived as hired goons and through haphazard underworld deals (Abdullah 1998; Zack-Williams 1995), the term rarray can also be deployed as a concept that explicates the "logic of straining" among Freetown's poor urban youth who devise various forms of "provisional agency and ... dynamic forms of waithood" (Finn and Oldfield 2015, 1). Through this logic of provisional life and waiting, we can understand rarrays as personifying the politics of perpetual half-hearted expediency of, often, youthful ex-combatants whose populations have occasionally bulged within post-conflict cities. Rarray citizenship in Monrovia can easily be understood through the politics of "use and dump" that pushed former fighters back to savage subsistence as they protested against the regime (Ukeje and Iwilade 2012, 346). Rarray can thus be deployed as a concept to explicate the representation of ex-combatants as "marginal souls ... deported to the margins" (Utas 2003, 231) in postwar Monrovia.

Reflecting on *Beasts of No Nation*, the character of Supreme Commander Goodblood, who excludes his fighters from political and economic benefits by building a political and economic enclave for himself, personifies the breakdown of postwar inclusion and his deification of commercial alliances (Reno 1997). The end of war does not guarantee sustainable peace (Harris 1999) but rather enables "social construction of difference between various groups" (Bøås and Utas 2014, 50). This is the context in which the rarrays, epitomizing postwar urban misfits, emerge. The idea that the streets and forests are indeed retreat sites for self-reorganization among disenfranchised ex-fighters phenomenally alters the way we perceive the fighters and challenge our assumptions about the child soldier. The question, who is the beast? is implicitly coeval with the indecisive way in which this film handles the subject of the rarray citizen. Goodblood's hostility to his fighters suggests that the city, in the illusory sense in which it promises social, political, and economic power in the aftermath of the war (Gates and Nordås 2010; Nagbe 1996, 53), often rescinds on this promise. Furthermore, the trenches, in the sense that they symbolize graves, draw on a well-established visual trope of death, which prevails even in the representations of Monrovia's streets, which are laden with bullet cartridges in the key films discussed in this chapter. The image of the streets littered with bullets is comparable to that of the trenches littered with fighters, a representation that urges us to see this film not just through the motif of violence but through the public life of the fighters who are themselves expendable capital of war. These ex-fighters embody a form of urban citizenship associated with deprived rights. In the subsections that follow, the chapter explores the representations of Monrovia's rarrays in cinema. It further explores two dimensions of the rarray's urban life that characterize their urban practices both during and after the war. These are, first, the idea of a necropolis as a place that normalizes the atrocities of war and which actuates expendability; and second, the social, political, and economic fractures as the underlying framework of urban chasms.

Historicizing Monrovia's rarray's in cinema

While the rarray today is a direct consequence of urban chasms, the term has a historical lineage. It originated from Sierra Leone where in as early as 1917 territorial gangs comprised of unemployed youths occupied various spaces

(O'Sullivan 2010). Subsequently, the term refers to the groups of marginalized urban youths, educated and uneducated, who mobilized in Freetown, Sierra Leone in the 1970s. In the beginning, they were considered criminals, "good-for-nothing people" (Abdullah 1998, 208). For Meghan Graham (2012, vi), these young toughs were "criminal, undereducated, unemployed and unemployable. They lack community and familial ties and are void of political sophistication … In this sense, they resemble the archetype of the African savage construct." Clearly, in the beginning, rarray denoted an urban surplus: comprised mainly of fighters who had nowhere to go after the end of the war, they thus congregated in cities with no firm social, economic, or political linkages. The term rarray boys was popularized as an urban identity only when hitherto criminalized groups' political vision acquired substantial appeal among the working class and university students. The term rarray is used here to capture expendability and marginality as an essence of urban identity – attributes that are not dependent on whether or not the rarrays are fighting or there is a war, but serve as a designation of people to be avoided at all costs (Abdullah 1998, 208). This isolation of the rarray within the city and its environs and the resulting metaphor of disconnection suggests a crisis of urban belonging in Monrovia.

Monrovia's recent history is undeniably upsetting. This is most notably a consequence of the prolonged civil war that ravaged Liberia. This kind of history has had great influence on the experiences of urban citizenship at the end of the war and in the kind of film narratives about Monrovia that have surfaced since. Furthermore, because Liberia has been mainly promoted as a setting for many globally known films rather than as a local production industry, both local and international filmmakers have explored Monrovia's history and aftermath of war in diverse ways. Such efforts are, however, founded on the effect of war in the city and the kinds of urban possibilities it has produced or motivated. The effect of civil war in Monrovia is twofold.

First, whereas the war displaced many and thus destabilized any possibility of organized film production in Liberia, it nevertheless became raw material for many film narratives about Monrovia. Governed through successive insurgences since Master Samuel Kanyon Doe of People's Redemption Council (PRC) led the assassination of Americo-Liberian President, William Tolbert of True Whig Party (TWP) in 1980, Monrovia has been a war city. This history of siege and bloody civil conflicts would continue with Dakhpannah Charles McArthur Ghankay Taylor's incursion in December 1989, subsequent capture and assassination of President Samuel Doe by Prince Yormie Johnson's combatants on 9 September 1990, and finally, Taylor's election to the presidency in July 1997. Taylor's later ousting in 2003 after Liberia's second civil war (Foster et al. 2009, 183–184) paved the way for the subsequent election of Ellen Johnson Sirleaf. The reality of mutilation, rape, death, starvation, disease, displacement, kidnappings, and other forms of torture and violence was commonplace both in life and the public imaginary of this time. Edward Zwick's *Blood Diamond* (2006), Garrett Batty's *Freetown* (2015), Greg Campbell's *Hondros* (2017), Gini Reticker's *Pray the*

Devil Back to Hell (2008), Andrew Niccol's *Lord of War*, and Sean Penn's *The Last Face* (2016) are some of the recent films that reflect on the verisimilitude of civil war among Liberia's population. Second, the end of the war brought with it a windfall in Liberia's film industry. While one would note, almost sadly, that the catalogue of films from Liberia, by Liberian filmmakers, is brief, the surge in homegrown Monrovia films – what Thomas Page (2017) calls the golden age of Liberia cinema – is partially attributed to the returnees who had been displaced by Liberia's Civil War of 1989–2003. Some of them, having acquired film production skills in Ghana and Nigeria, used these skills back in Liberia to boost the industry. In the same breath, some notable cross-border film productions have focused on Liberia, for example, Mykel C. Ajaere's *Liberian Girl* (2011) and later Courage Borbor's *Hatred* (2012). However, such productions have not been seen as successfully achieving a verisimilitude of the Liberian narratives they sought to tell but rather as part of Nollywood's migrant archives (Noah 2019, 278).

Besides war, the other factor that has influenced filmmaking in Liberia is the impromptu market generated by the 2014 Ebola outbreak in the city. Using the moniker "Ebolawood," Page (2017) discusses the role of closing Liberia's boundaries, which barred importation of pirated films, boosting the work of local filmmakers. Among the beneficiaries of this windfall was Richard Dwumoh, an outlier producer credited with the production of various locally acclaimed films under his private production house. These films, most of them comedies, are the flag-bearers of Liberia's film industry, also christened Lollywood. Yet, this golden age was short-lived, ending with the 2016 lift of import bans after the end of Ebola pandemic. Other popular Liberian films adopt a quasi-documentary approach to narrate about social issues such as Ebola (see Mitman and Siegel's *In the Shadow of Ebola* (2014) and Darg's *Body Team 12* (2015), for instance); effects of war (such as Vale's *Small Small Thing* (2013) and Brabazon and Stack's *Liberia: An Uncivil War* (2004)). The award-winning Seema Mathur's *Camp 72* (2015) is perhaps the most recognizable film dealing with the postwar reconciliation efforts in Liberia.

Yet, locally produced Monrovia films have had noticeable limitations in representing urban peculiarities in an enduring or globally noticeable scale. This happened on a number of levels. First, local films addressing urban pre-carity, which peaked during the period of Ebola pandemic, dwindled soon after. Second, such films tended to have less international reach, mostly circulating within local or regional markets. Accordingly, they do not historicize the larger concerns, such as postwar crisis urbanism in Monrovia or the superficies of postwar crisis urbanism in the context of Monrovia's recent history. This absence surfaces especially in Liberia's war films directed by foreigners, such as *Johnny Mad Dog* (2008) and *Lord of War* (2005), which overtly engage everyday street life and urban citizens.

Directed by French filmmaker, Jean-Stephane Sauvaire, *Johnny Mad Dog* is an adaptation of Emmanuel Dongala's novel of the same name (*Johnny Chien Méchant* 2002), which tells the story of Liberians United for Reconciliation and Democracy (LURD) child soldiers led by *Johnny Mad*

Dog (Christophe Minie), a teenage boy, during Second Liberian Civil War, that led to the ousting of Charles Taylor in 2003. The novel features dual protagonists: a 16-year-old girl called Laokolé, who embodies the psychological effect of the physical displacement of citizens in civil war, and *Johnny Mad Dog*, a teenage child soldier, who symbolizes social and economic displacement in wartime Monrovia. The novel has been described by Publishers Weekly (2005) as an:

> Unflinching look at the greed and ignorance that drives fighters like Mad Dog, as well as the fear, desperation and anger of those trapped in the cross fire … (and) frames some powerful questions: namely, how humans can be so cruel and conversely, how do they maintain their humanity in the face of unremitting ugliness?

This description reiterates the heuristic qualities of violence from cause (greed and ignorance) to effect (fear, desperation, and anger), which eventually dominate the visual style of the film in its depiction of Monrovia, where most of the story is set. Yet, when we pay attention to the juxtaposition of Mad Dog's militant character with the civilian character of Fatmata – Lovelita (Careen Moore), whom he kidnaps as his sex slave, we can begin to explore a range of underlying collective issues such as vulnerability, anxiety, and uncertainty for the fighters and the civilians. Further, through the character of Laokolé (Daisy Victoria Vandy) – a young girl of the same age as Johnny and Fatimata – who is looking after her ailing father and younger brother as war rages on, we get immersed in the mental world of wartime Monrovia, where the physical is entwined with the psychological. The inability to occupy the city as a civilian resident or to commandeer it as a child fighter spells out the uncertainty with which armed fighters and civilians inhabited Monrovia. Thus, what the film achieves by following a group of child soldiers marching across the countryside and later through the city and juxtaposing their violent aggressions with the everyday suffering of civilians is the conceptualization of a continually disoriented and destabilized sense of urban life. As surmised in Peter Bradshaw's (2009) review of the film, where he terms child soldiers as the "evolutionary endpoint of war" and refers to war as "brutalizing, infantilizing, dehumanizing, requiring the unquestioning submission to authority," *Johnny Mad Dog* may be understood as a film about the effacement of normal urban life in moments of war. This narrative tableau is also central in *Lord of War*.

Andrew Niccol is a New Zealand-born screenwriter-director based in America. Prior to making *Lord of War*, he had directed several internationally acclaimed films such as *Gattaca* (1997) and *Simone* (2002). Afterward, he directed other films including *In Time* (2011), *The Host* (2013), *Good Kill* (2014), and *Anon* (2018). In *Lord of War*, he narrates about Yuri Orlov (Nicolas Cage), an American immigrant legally registered as a Jew, but originally a Russian, who runs illegal arms deals with Andre Baptiste Sr. (Eamonn Walker), a Liberian warlord emblematic of endemic corruption and abuse of power within Monrovia's top leadership (Osaghae 1996). His militant actions positing war as a necessary façade for looting the country's mineral resources, he personifies the philandering or cannibalism of the country's resources (Reno 1996, 1997). Buying arms from post-Cold-War armouries around the world, Yuri supplies them to war zones in Africa, participating in a network of foreign military armoury bosses, arms dealers, arms traffickers, corrupt military forces, and corrupt African governments. The narrative focuses on the personal experiences of violence resulting from Yuri's arms supplies to Liberian warlords. It thus registers the convergence of global arms players with Monrovia's violent history and politics of urban citizenship. Niccol's film is thus grounded in the accoutrements of Monrovia's urban crisis, typified foremost by random violence. In recent reviews of the film, Rahul Hamid (2006, 53) comments on the circuit of violence emanating from America to devastate the world and the role of America's arms industries in African politics, while Kim Newman (2000) focuses on the fetishism around illegal gunrunning. Beyond the centrality of military globalism in the film, what is most peculiar about the film's representation of Monrovia is the expendability of the city's residents not only during but also after the war (Hamid 2006).

Both *Johnny Mad Dog* and *Lord of War* are built around cyclic violence in west Africa. Both films use a quasi-ethnographic approach to achieve a verisimilitude of war, and hence anchor their narratives within Monrovia's contemporary history of civil war, which ended in 2003. By historicizing the city, they reflect on the potential effect of war on both the urban form and urban practices. It is noteworthy that the films are told from the perpetrator's viewpoint, positioning the procurement of violence as a key metric of Monrovia's urban life. In other words, the city is reduced to a symbol of wanton aggression, and its afterlife precariously attenuated through the figure of the fighter who is displaced both from the urban commune and from any inclusive endpoint. Accordingly, both films' narratives advance through the iconicity of chasms, both in literal disconnection between urban communities, and in the psychological rifts that fuelled the war. Chasms here refer to the grouping of urban residents into inequality-creating identities. It is also used as a concept that articulates the experiences of disconnection and hierarchic existence in the aftermath of war. Chasms can be construed through individual aggrandizement, which tends to marginalize ex-fighters in postwar Monrovia in a manner even greater than it does for the citizens displaced during the war. Paying attention to expendability as a prevalent narrative motif in these films, the remainder of this chapter discusses the representations of Monrovia's postcolonial urban life in the context of excluded citizenship characterized by politically and economically displaced citizens. These films, it is argued, construe a discourse of a city whose residents remain peripheral to its progress, and are hence "stranded." Using the term "rarray citizenship" to refer to the crisis personified by such urban experiences, I advance my discussions on two fronts: first, the idea of a necropolis city and second, the idea of urban fractures.

Necropolis

Achille Mbembe (2001, 199) describes the life of the postcolonial subject as follows:

> Enclosed in an impossibility and confined on the other side of the world, the natives no longer expect anything from the future. A time has got farther away, leaving behind only a field of ruins, an immense weariness, an infinite distress and a need for vengeance and rest. This nameless eclipse is also accompanied by a proliferation of metaphysics of sorrow, of thoughts of final things and days. The proliferation is partly due to the excessive burden of mass suffering and the omnipresence of death. Dying, often prematurely, for nothing, no apparent reason, just like that, without having sought death, constitutes the soil of recent memory. Through the brutality and uncertainties of everyday existence, the fear of dying and being buried has also become the way the future, inexhaustible and infinite, is foreshortened and accomplished.

For Mbembe, the postcolonial subject's everyday life is not only unproductive but characterized by inexhaustible anxiety. Further, the thought of death as inseparable from the very existence of the postcolonial subject draws attention to the significance of "the imminent" and "the eventual" as very much part of everyday anxieties. Death is, in Mbembe's view, either the physical form, where one may just find themselves in violence and starvation, or diseased and dying, "without having sought death" (Ibid.). Death is also a condition of hopelessness in life – where subjects are bound in impossibility, with no future hope and living in present circumstances that are very difficult. These views echo citizenship experiences in Monrovia where The Advocates for Human Rights (2009) reports extensive tactics of violence and the desperate conditions in which many Monrovia residents lived during and after the war. Here, citizens trapped within violence that they had nothing to do with, would find themselves casualties – many having died, again, "without having sought death" (Mbembe 2001, 199). The wantonness of death, so typical of wartime, is a useful perspective to critique the valence of violence as the initial symbol of the expendability of civilians and fighters and the necropolis as a normalization of the atrocities of war that actuate such expendability. With this framing of the post-colony, Mbembe provides a conceptual framework with which I critique representations of Monrovia as a necropolis.

Lord of War starts with a montage showing the lifecycle of a bullet from its manufacture in a Russian factory, through shipping to rebel fighters in Monrovia, and eventual firing out of a gun muzzle and penetration of the head of a young African boy in the streets of Monrovia. This final shot is framed from within the barrel of the gun, thus adopting the perspective of the killer bullet. The montage is framed from the point of view of a bullet, which is positioned at the horizontal centre of the frame, giving it dominance. That the camera shows the bullet's point of view from inside the barrel to the boy's forehead assigns culpability to that specific bullet from the Russian factory now weaponized in an African wartime city. As the bullet hits the boy, the shot is truncated by a swift cut-to-dark transition, mimicking a rapid blink, maybe even a reflex to obfuscate the carnage it represents. Here, the idea of a death that has arrived without being sought, and Monrovia as the place of such occurrence, is implicit. This opening montage characterizes the omnipresence of death in everyday life in Monrovia, aptly supporting the city's symbolism of a necropolis.

Yet, beyond conveying the unpredictability of carnage as part of everyday urban life in Monrovia, the montage also expresses the expendability of city residents as part of everyday street life. Such expendability may also be read as a trivialization of death which, at the height of Monrovia's factional wars, is well articulated. Foster et al. (2009, 141–142) describe how "Prince Johnson reportedly sat in a chair on top of a table, playing a guitar and singing, while his soldiers randomly killed people" inside an Economic Community of West African States Monitoring Group (ECOMOG) ship, where refugees had sought protection after the assassination of Samuel Doe. Sanctioned by a rebel faction leader who fought Doe's regime because of its atrocities, this is a paradoxical massacre. In *Lord of War*, the most compelling signal of how close death and life are linked in the streets of Monrovia – hence, the trivialization of death, which also remains vivid throughout almost the entire film – can be grasped in two scenes.

The first scene occurs early in the film, and shows a vulture next to a lone corpse, juxtaposed with Yuri and Andre Baptiste Jr., chatting unaffected in front of a decrepit hotel. In this shot's foreground, a vulture stands atop a corpse lying on the ground while a group of people go on with their business next to an abandoned building, which dominates the whole background, undisturbed. Some people are armed while others appear to be civilians. There are two bullet cartridges besides the corpse. The initial impression of the shot is both the normalcy of death in everyday life in Monrovia and the anxieties of civilian life surrounded by armed groups. Andre Baptiste Jr., Yuri, and the other characters in the shot seem unbothered by the vulture devouring the body in front of them. Indeed, the cinematography here uses shadows and light to demarcate the foreground (with the corpse) and the background (with the living).

Thus, the most compelling reading of this shot is not that there is rampant death in the streets, but that life goes on, and that no one seems to notice or is bothered by this death. The characters in the background are visually isolated from the corpse in the foreground by the different lighting, which suggests this division of existence. Further, the framing of the corpse in relation to the initial position of the killer bullet in the opening montage of the film is notable. The corpse occupies the vertical centre of the screen, the same visual position as the bullet, thus providing a visual continuity to the centrality of death and shifting attention from the Russian factory's involvement in Monrovia's bloodshed, to the normalcy of this death in this city. If we read this framing as a continuity of the initial sequence where a young boy is fatally shot, then this shot signposts Monrovia's streets as "the scene of the crime" (Seltzer 2003, 62), alluded to in the opening montage. In this regard, the composite lighting conveys the simultaneity of death and life in the streets of Monrovia, and hence the precarity of life. Consequently, positioning the corpse in the foreground may be read as a signal that the film's story of Monrovia is foremost about the expendability of human life. This reading is supported by the use of cinematographic choices positioning the killer bullet and its victim centre screen, emphasizing the centrality of this image in Monrovia's postwar imaginary. It is also supported by the mise-en-scène of Yuri, who appears in a swathe of spent bullet cartridges as he supplies weapons to the fighters. The shot's mise-en-scène of armed soldiers positioned at various points around the hotel further attests to the centrality and immediacy of violence in the city, and death, as embodied by the civilian corpse abandoned in the foreground. Both aspects accentuate the centrality of carnage in Monrovia's everyday life and also allow the viewer to discover and acclimatize to bloodshed in the city. This cinematographic style and mise-en-scène of death and violence constitute a metaphor for the expendability of human life amid Monrovia's violence. This montage of a corpse cannot be read in isolation of the wider scope of carnage, as it encapsulates just a fraction of Monrovia's topography of death in this film. If the notion of one bullet, one victim, and one corpse – which is set out in the opening sequence of the film – map out death in this city, the film's subsequent scenes allude to the vastness and contiguity of this carnage.

The second scene occurs later in the film and shows Yuri standing in a street littered with bullet cartridges. This scene opens with a fade-in transition from a black screen as the camera dollies above vast tracts of bullet cartridges strewn about the street, finally revealing Yuri in a deserted street occupying the vertical centre of the screen – the same position in the frame as that occupied by the vulture. After the initial high-angle craning shot where the camera hovers closely above the bullet cartridges, the camera angle changes from high to low just as it dollies to Yuri standing in the middle of the shot. This craning movement prolongs the exposure time of the empty shells, subtly emphasizing the enormity of Monrovia's war. The background is filled with smoke and wires hang loosely from electricity poles. This scene uses the mise-en-scène of fire and plumes of smoke to highlight the presentness of war and carnage. In the specific case of Liberia, this imagery of bullet cartridges coupled with the empty streets documents the cycle of bloodshed and invokes death as emblematic of the city's postcolonial crisis. Here, carnage perpetrated in pursuit of economic predation not only leads to political expediency (Atkinson 1997), but also necessitates an urbanism of flight from an ever-present risk of expendability. Such a claim is made cautiously here, as this evocative scene of carnage and Yuri's dominance in the shot may as well suggest global interference with Monrovia's political conversations, and the wanton use of violence for greedy geopolitical reasons rather than redemptive purposes.

Death in these two scenes is a crypt for the crisis of urban life and hence eulogizes Monrovia's perpetual limbo between conflicts whose beginnings and endings are uncertain. Here, Monrovia's essence is characterized by endless trauma (Waugh 2011). In the images of nearly deserted streets we see violence as the prominent dimension of Monrovia's urban life and the city's residents as an afterthought in the general political narrative unfolding in the film. This contiguity of violence and displacement may thus illustrate the historical dilemma of politics in Liberia and, specifically, the hideousness of violence in Monrovia. Since its founding as a settlement for freed black American slaves in 1847 (Duyvesteyn 2005, 21), the Republic of Liberia has been ruled through a series of what Christensen and Utas (2008) term "polit-ricks" of governance, randomly effected through factional armed dissident groups and a fragile central regime.

The history of the nation is tied to this history of disenfranchising politics that do not address Liberians' most pressing "issues of concern" (Söderström 2015, 158). In this context, death in this film becomes an allegory of Monrovia's transition from the exploitative and alienating Americo-Liberian leadership to the gangster-like leadership where victimization of citizens is part of the process of political concessions. Recent civil wars – most notably Taylor's invasion from 24 December 1989 to 29 November 1990, then again from 15 October 1992 to 31 July 1993, and even his election to the presidency in 1997 – may be indicative of how war (and its inevitable consequences such as death and displacement) in Liberia usually assumes the place of a political conversation. Further, LURD's and Movement for Democracy in Liberia's (MODEL) armed invasion of 2003, which led to Taylor's resignation, signalled the elusiveness of peace (Duyvesteyn 2005, 27). The consequence of this constant reality of invasion between 1990 and 2003 was that Monrovia was habituated to "short-lived and uneasy peaces ... [and] lulls inaugurated by cease-fires, accords and agreements" (Outram 1997, 355). Against this background, the representations of Monrovia's largely absent urban residents appear as signifiers of civilians' exclusion from the (benefits of the postwar) city, and hence signify the crisis of expendable life as the very form of the city's postwar urbanism.

What is achieved by such an image of a vacant Monrovia is "dismissing the [city's] human dimension" (Blanchard 1985, 118), not because it is inconsequential to understanding Monrovia, but because the street on which the film is set is mainly identified as a place of mass displacements and carnage. The empty cartridges littering the surface are metonymic of miniature graves, hence allude to the colossal loss of life in the city. The idea of necropolis shaped by the images of carnage, it can be argued, is the means through which the film cultivates a sense of crisis urbanism. Among the consequences of this crisis are social, political, and economic fractures. Catalysed by the economies of war and ethnic allegiances, these chasms are explored in depth in *Johnny Mad Dog*.

Fractures

The metaphor of chasms in the context of Monrovia invokes the political, social, and economic trauma carried over from years of carnage, insurgency, and civil war. When these postwar narratives are relayed in *Johnny Mad Dog*, the contrasting and divergent prospects that political personages and ex-combatants faced in postwar Monrovia are at the fore. In this film, this divergence posits spatial detachment as emblematic of political detachments. Thus, whereas *Johnny Mad Dog's* entry to Monrovia – fully garbed in a wedding dress and armed with a rifle – is portrayed as a symbolic journey into political, ethnic, and economic inclusion, the film's ending intimates a contiguity of betrayal and dispossession, thus presaging the disconnections between places, ethnicities, political groups, and socio-economic classes. These incidental chasms are foremost signified by the image of a broken bridge, which dominates the film's initial stages.

In *Johnny Mad Dog*, the mise-en-scène of the bridge dominating the shot where the fighters are about to breach the government defences guarding the city is comprised of a waterway in the foreground, a broken bridge and a sprawling shanty neighbourhood in the middle ground, and the Atlantic Ocean in the background. The waterway extends from the foreground separating the slum-like settlement from the mainland, suggesting the settlement's isolation by the waters of the Atlantic Ocean. The composition and wide-angle framing give context to this detachment between the tiny settlement and the mainland, creating an aura of desolation and vulnerability, and hinting at the settlement's precarious subsistence. Additionally, the wide framing allows the viewer to contemplate the two spaces previously connected by the bridge, and thus become aware of their current state of disconnection. That the shot is framed from a wide angle and hides closer details from the viewers metonymically references a city that conceals its possible social, economic, and political fissures from a casual glance. The dominant wide shot of the broken bridge is indicative of a concealed view, of a broken city, and of the ideological and political fissures that have seen Monrovia, the ultimate prize in the various political conflicts, divided and administered through various ceasefire deals (Foster et al. 2009, 165). Yet, attention is drawn to this symbolism of a chasm between the mainland and the vast slum by the use of wide shot. This cinematographic style speaks of Monrovia's history, in which bridges were easily seized as defensive positions, as happened in the 2003 combat between Charles Taylor's troops and the LURD rebels who sieged Monrovia. Here, bridges acted as frontiers between urban spaces controlled by warring factions, thus as de facto boundaries between urban enclaves. One way to seize bridges was barricading with razor wire that cut off access to either the government or rebel sections of the city (CNN 2003). Here, the bridge prohibited access and participation in political and economic prospects within opposing territories. Accordingly, the image of a broken bridge, which dominates the film's initial shots of Monrovia in *Johnny Mad Dog*, makes a significant contribution to the interpretation of urban chasms.

In the long wide shot when we first see a bridge in *Johnny Mad Dog*, it is visually eclipsed by the mise-en-scène of surrounding water. The vast body of water reinvigorates within the viewer's mind the significance of the Atlantic Ocean in the overall semiosis of this shot. Particularly, it refreshes Monrovia's history of invasion, which started when Captain R. F. Stockton, the leader of the Americo-Liberian settlers moored along the Atlantic coast, ordered King Peter of the Dey and the natives of Montserrado to surrender territory at gunpoint on 15 December 1821 (Waugh 2011, 16). Incidentally, the Atlantic Ocean here reinvigorates the memory of the Americo-Liberian invasion, which persisted until Charles Taylor's resignation from the presidency, and hence signposts the history of disconnection between tribes and ethnicities, which, in Monrovia, have been most clearly actualized in the spatial divisions of the city claimed by opposing factions during times of war. Such intrinsic historical connotations can then be read in the context of disruption premised by this shot's geometric figure comprised of the foreground-background and the left-right axes formed by the intersection between the bridge and the river.

The foreground-background axis comprised of the river connecting with the ocean beyond the bridge emblematizes the invasiveness of the channel connecting the ocean, the signifier of western intrusion, to the interior. Here, the resulting symbolism of arrival and penetration of the city by Americo-Liberians – who viewed themselves as superior to the local tribes over whom they imposed their political leadership – signals a disruption of social, economic, and political order. Further, the river flowing into the ocean, literally emptying into the Atlantic, is a channel through which resources could be siphoned from inland and heading west. The shot thus easily symbolizes plunder of resources, and by extension, the reinforcement of inequality in wartime Monrovia's highly exploitative capitalist economy. This reflexive representation of the city suggests that, founded at gunpoint, Monrovia cannot be represented in isolation from the interruption arising from Captain R. F. Stockton's and King Peter of the Dey's deal, which not only brought freed slaves into Monrovia (the river as inlet) but also seized their freedom and resources at the same time (river as outlet). *Johnny Mad Dog* echoes these reverberations of the city's history of intrusion and overthrow, particularly so the ethnic and political chasms that have shaped urbanism in Monrovia.

The left-right axis comprises the broken bridge, which would otherwise connect two sides of the city. Here, the broken bridge signifies a disconnect between the lagoon settlement and the rest of the city, and thus hints of enclaves. The broken bridge is indeed a physical barrier presaging the compartmentalization of Monrovia's communities; hence, it enables the viewer to witness the heightened ethnic and political compartmentalization of the city through the battles between the various factions across the bridge. In this respect, the broken bridge is an impression of broken promises, broken pacts, broken expectations, and the fake coalitions between rebel factions that culminate in quick political fallout (Duyvesteyn 2005, 25–27; Foster et al. 2009, 163). It embodies a perpetual crisis of a "broken" Monrovia whose communities are irreparably kept apart, again replenishing historical memory with failed political deals, such as that between President Doe, ECOMOG soldiers, and Prince Johnson's Independent National Patriotic Front of Liberia (INPFL) rebels, who collectively guarded the city after Taylor's attack (Duyvesteyn 2005, 30). The stated objective of this coalition was not to safeguard Doe's power, but to prevent their mutual opponent, Charles Taylor, from ascending to power. Soon after, President Samuel Doe was tortured to death by the same INPFL fighters (Waugh 2011, 149–152), clearly indicating the political and ethnic chasms that have characterized Monrovia's streetscape for a long time. In this case, the notion of disconnection assumes an immediate political connotation of detachment from the political cause embodied by the characters fighting to seize the bridge. Within *Johnny Mad Dog*, such compartmentalization indexes the "political and ethnic cleavages that continue to haunt Liberia" (Bøås and Utas 2014, 50).

Monrovia's ethnic conflicts are inherited from economic, political, and ethnic enclaves typical of hinterland politics. For instance, the 1985 Doe-Quiwonkpa ethnic animosity resulted in opposed tribal and political enclaves, with Krahn and Mandingo on one side, and Gio and Mano on the other. Charles Taylor's subsequent uprising against Doe's government appealed to the oppressed and harassed Gio and Mano ethnic tribes of Nimba County. His annexation of Greater Liberia, also known as "Taylorland," and establishment of a rival National Patriotic Reconstruction Assembly Government (NPRAG) administration and capital city at Gbarnga in Bong County were directly enabled by and in turn promoted these enclaves. Thus, while the annexation involved reserving physical territory from Doe's government to signal the establishment of competing political centres, it also intimated the centrality of ethnic alliances in solidifying political groupings. A subsequent decisive battle in this political feud was staged in Monrovia, the symbol of ultimate political power. When read within this history, the film's broken bridge invokes Liberian divisions that have mutated from rebel-government antagonisms to ethnic rivalry, and the more enduring divisions of the city into competing interests, ethnicities, political groupings, rebel factions, and habitation zones. In *Johnny Mad Dog*, the severity of these ephemeral territories is articulated in a subsequent scene set on the surface of the bridge, with mise-en-scène of the razor wire on the foreground and explosions in the middle ground.

This medium shot reveals more grotesque details of a city beleaguered by extreme violence. The razor wire laid on its surface is metonymic of urban spatial demarcations, while the shoe dangling in the razor wire gives agency to this scene by suggesting hasty flight. In the context of this film where characters are in constant flight from the violent rebel fighters, such composition draws an anecdotal relationship between the emptiness of the used bullet cartridges strewn on the surface of the bridge and the sense of loss that most characters in the film feel. The metaphor of laceration, curated by pieces of what are perhaps torn clothes or packages dangling from the razor wire, enhances the perception of violent separation. This motif usefully caricatures the violent separation of Monrovia's urban communities from each other, which are often actuated through revenge attacks, kidnapping, cannibalism, political aggression, or occupation. It is thus arguable that the idea of territorial control and the possibility of siege suggested by this razor wire reconstitute the dialectic of violent demarcation of the city into disconnected communities.

Analogously, the explosion and bullet cartridges emphasize the severity of violence and amplify how the ambience of war controls Monrovia. That is, throughout this film, the whirring and ricocheting bullets, screams, exploding bombs, delirious shouts, weeping sounds, crying babies, and even silence make up Monrovia's aural landscape. The derisive mood that Sauvaire cultivates through this fusion of warzone sounds and images of extreme violence can be interpreted as a narrative strategy to enshrine Liberia's contiguous quasi-liberation wars within the enclaves upon which the country's many civil wars have been fought. The rebel-government conflict alluded to here usefully signifies a political chasm between the rebels and the incumbent political leadership of the Dogo ethnic group. Taking the place of urban communities, the bullet cartridges aptly express Monrovia's streets as spaces for "wound ... violence and loss ... [and] the city's scarred identity as wound landscape" (Seltzer 2003, 62). The bullet cartridges, emblematic of once-portentous weapons of war now existing as debris in the postwar city, actuate the symbol of fracture in postwar Monrovia, namely, the politics of "use and dump" (Ukeje and Iwilade 2012, 346) as the ultimate form of urban fracture, whereby rebel fighters are abandoned as spent "war" debris, hence irrelevant. They caricature the city's post-conflict aversion to the disbanded ex-combatants trying to "escape the city they once sought to control" (Hoffman 2007, 401), now relegated to a predicament of political debris. This fracture between ex-fighters and the political figures they fought for, and their resulting abandonment, provide grounding to discuss how the character of rarray boys, mostly rendered redundant by the end of the civil war in Monrovia, represent a dominant form of urbanism in postwar Monrovia.

Bibliography

Abdullah, Ibrahim. 1998. "Bush Path to Destruction: The Origin and Character of the Revolutionary United Front/Sierra Leone." *Journal of Modern African Studies* 36 (June): 203–235.

Ambe, Hilarious N. 2004. "Shit and Stench as Dramatic Strategy: Bate Besong's Beasts of No Nation." In *The Literary Criterion,* edited by C. N. Srinath et al., Vol. xxxix (3 & 4): 185–197.

Atkinson, Philippa. 1997. *The War Economy in Liberia: A Political Analysis.* London: Overseas Development Institute.

Blanchard, Marc Eli. 1985. *In Search of the City: Engels, Baudelaire, Rimbaud.* Saratoga, CA: Anma Libri.

Bøås, Morten, and Mats Utas. 2014. "The Political Landscape of Postwar Liberia: Reflections on National Reconciliation and Elections." *Africa Today* 60 (4) (Summer 2014): 47–65.

Bradshaw, Peter. 2009. "Johnny Mad Dog." *The Guardian,* October 22, 2009. https://www.theguardian.com/film/2009/oct/22/johnny-mad-dog-review.

Christensen, Maya M., and Mats Utas. 2008. "Mercenaries of Democracy: The 'Politricks' Of Remobilized Combatants in the 2007 General Elections, Sierra Leone." *African Affairs* 107 (429): 515–539.

CNN. 2003. "Liberians Rush Bridges to Port." *CNN,* August 15, 2003. http://edition.cnn.com/2003/WORLD/africa/08/15/liberia0635/.

Dongala, Emmanuel. 2002. *Johnny Chien Méchant.* Paris: Le Serpent á Plumes.

Duyvesteyn, Isabelle. 2005. *Clausewitz and African War: Politics and Strategy in Liberia and Somalia*. London and New York: Frank Cass.

Finn, Brandon, and Oldfield, Sophie. 2015. "Straining: Young Men Working Through Waithood in Freetown, Sierra Leone." *Africa Spectrum* 50 (3): 29–48.

Foster, Dulce, Dianne Heins, Mark Kalla, Michele Garnett McKenzie, James O'Neal, Rosalyn Park, Robin Phillips, Jennifer Prestholdt, Ahmed K. Sirleaf II, and Laura A. Young. 2009. *A House with Two Rooms: Final Report of the Truth and Reconciliation Commission of Liberia Diaspora Project*. DRI Press Saint Paul: The Advocates for Human Rights.

Gates, Scott, and Ragnhild Nordås. 2010. *"Recruitment and Retention in Rebel Groups."* Paper presented at *the Annual Meeting of the American Political Science Association*, 2–5 September, 2010, Washington, DC.

Graham, Meghan Elizabeth. 2012. *Rarray Boys to the Savis Men: The Young Turfs of Sierra Leone*. San Diego, CA: San Diego University.

Hamid, Rahul. 2006. "Review of Lord of War." *Cinéaste* 31 (2): 52–55.

Harris, David. 1999. "From 'Warlord' to 'Democratic' President: How Charles Taylor Won the 1997 Liberian Elections." *The Journal of Modern African Studies* 37 (3) (September 1999): 431–455.

Hoffman, Danny. 2007. "The City as Barracks: Freetown, Monrovia, and the Organization of Violence in Postcolonial African Cities." *Cultural Anthropology* 22 (3): 400–428.

Hoffman, Danny. 2011. *The War Machines: Young Men and Violence in Sierra Leone and Liberia*. Durham: Duke University Press.

Käihkö, Ilmari. 2018. "Constructing War in West Africa (and beyond)." *Comparative Strategy* 37 (5): 485–501.

Mbembe, Achille. 2001. *On the Postcolony*. Berkeley/Los Angeles/London: University of California Press.

Nagbe, Moses, 1996. *Bulk Challenge: The Story of 4,000 Liberians in Search of a Refuge*. Cape Coast: Champion Publishers.

Newman, Kim. 2000. "Lord of War Review." *Empire Online*, January 1, 2000. https://www.empireonline.com/movies/lord-war/review/.

Noah, Tsika. 2019. "Nollywood Chronicles: Migrant Archives, Media Archeology, and the Itineraries of Taste." In *A Companion to African Cinema*, edited by Kenneth W. Harrow and Carmela Garritano, 269–290. Hoboken: John Wiley & Sons.

O'Sullivan, Tadhg. 2010. "Freetown's Freewheeling Graffiti." *The World*, January 22. https://www.pri.org/stories/2010-01-22/freetowns-freewheeling-graffiti.

Osaghae, Eghosa. 1996. *Ethnicity, Class, and the Struggle for State Power in Liberia*. Dakar: CODESRIA.

Outram, Quentin. 1997. "'It's Terminal Either Way': An Analysis of Armed Conflict in Liberia, 1989–1996." *Review of African Political Economy* 24 (73): 355–371.

Page, Thomas. 2017. "Ebolawood: Liberia's Golden Age of Film." *CNN*, August, 2017. https://www.cnn.com/interactive/2017/08/world/ebola-lollywood-liberia-film/.

Podder, Sukanya. 2011. "Child Soldier Recruitment in the Liberian Civil Wars: Individual Motivations and Rebel Group Tactics." In *Child Soldiers: From Recruitment to Reintegration*, edited by Alpaslan Özerdem and Sukanya Podder, 50–75. Hampshire: Palgrave Macmillan.

Publishers Weekly. 2005. *Johnny Mad Dog*. January 5. https://www.publishersweekly.com/978-0-374-17995-3.

Reno, Will. 2007. "African Rebels and the Citizenship Question." In *Making Nations, Creating Strangers: States and Citizenship in Africa*, edited by Sara Dorman, Daniel Hammett, and Paul Nugent, 221–240. Leiden and Boston: Brill.

Reno, William. 1996. "The Business of War in Liberia." *Current History* May 1996: 211–215.

Reno, William. 1997. "African Weak State: Survival and New Commercial Alliances." *African Affairs* 96 (383) (April): 165–185.

Seltzer, Mark. 2003. "Berlin 2000: 'The Image of an Empty Place'." In *After-Images of the City*, edited by Joan Ramon Resina and Dieter Ingenschay, 61–74. Ithaca and London: Cornell University Press.

Simone, AbdouMaliq. 2011. *City Life from Jakarta to Dakar: Movements at the Crossroads*. London: Routledge.

Söderström, Johanna. 2015. *Peacebuilding and Ex-Combatants: Political reintegration in Liberia*. London and New York: Routledge.

The Advocates for Human Rights. 2009. *CEDAW - 44th Session - July 2009*. Comments by the Advocates for Human Rights on the Combined Initial, Second, Third, Fourth, Fifth, and Sixth Periodic Reports Submitted by Liberia, Minnesota: The Advocates for Human Rights.

Ukeje, Charles Ugochukwu, and Akin Iwilade. 2012. "A Farewell to Innocence? African Youth and Violence in the Twenty-First Century." *International Journal of Conflict and Violence* 6 (2): 339–351.

Utas, Mats. 2003. *Sweet Battlefields: Youth and the Liberian Civil War*. Uppsala: Department of Cultural Anthropology and Ethnology, Uppsala University.

Waugh, Colin M. 2011. *Charles Taylor and Liberia: Ambition and Atrocity in Africa's Lone Star State*. London and New York: Zed Books Ltd.

Zack-Williams, Alfred. 1995. *Tributors, Supporters and Merchant Capital*. Brookfield, VT: Ashgate Publishing.

Filmography

Anon. Dir. Andrew Niccol, USA, 2018.

Beasts of No Nation. Dir. Cary Joji Fukunaga, Japan, 2015.

Blood Diamond. Dir. Edward Zwick, USA, 2006.

Body Team 12. Dir. David Darg, USA, 2015.

Camp 72. Dir. Seema Mathur, USA, 2015.

Gattaca. Dir. Andrew Niccol, USA, 1997.

Good Kill. Dir. Andrew Niccol, USA, 2014.

Freetown. Dir. Garrett Batty, USA, 2015.

Hatred. Dir. Courage Borbor, Nigeria, 2012.

Hondros. Dir. Greg Campbell, USA, 2017.

In the Shadow of Ebola. Dir. Gregg Mitman and Sarita Siegel, Liberia, 2014.

In Time. Dir. Andrew Niccol, USA, 2011.

Johnny Mad Dog. Dir. Stephane Sauvaire, France, 2008.

Liberia: An Uncivil War. Dir. James Brabazon and Jonathan Stack, Liberia, 2004.

Liberian Girl. Dir. Mykel C. Ajaere, Nigeria, 2011.

Lord of War. Dir. Andrew Niccol, USA, 2005.

Pray the Devil Back to Hell. Dir. Gini Reticker, USA, 2008.

Simone. Dir. Andrew Niccol, USA, 2002.

Small Thing. Dir. Jessica Vale, USA, 2013.

The Host. Dir. Andrew Niccol, USA, 2013.

The Last Face. Dir. Sean Penn, Spain, 2016.

6 Revolt urbanism

Cairo's crisis citizenship under construction

This chapter discusses the crisis of urban citizenship in Cairo. When the so-called Arab Spring finally erupted in Egypt, it not only premiered new approaches to negotiating and shaping governance but also marked a unique response to various pressures of citizenship. One major realization was that looking past an outlook of civility and obedience to governance, a seething rage had accrued. Taking that insight as a starting point, this chapter discusses Cairo's crisis citizenship as a venerable tug-of-war between latent oppressive manoeuvres and covert rebellion, a scenario that has produced an urbanism of revolt in cities like Cairo, where the effects of this struggle were most evident. Focusing on the representations of this emergent urbanism in Khaled Marie's *Asal Aswad* (*Molasses*) (2010), this chapter builds a theoretical connection between the overt street protests that erupted in Cairo in 2010 and the covert disenfranchisement that existed long before. A survey of post-2010 films and media reports giving a clear emphasis on the explicit crisis are the main material for this theorization. The chapter's main argument is that these urban film narratives convey an urban crisis in the form of latent oppression and resistance in everyday life in Cairo. This covert crisis, more than its climax as the Egyptian revolution, conveys the larger underlying urban crisis and thus provides a useful window into the meaning of urban citizenship in postcolonial Cairo. The chapter works across a range of symbols such as immigration, passports, transport, freedom, and state services as the frontiers of this crisis.

Crisis citizenship

The Egyptian uprisings of 2011 and 2013, generally classified under the Arab Spring, draw heavily on political economies of urban space and public performance of urban activism (Abaza 2014). The uprisings constituted a momentous occurrence that publicized the changing ideological and political orientation of the Arab region using the streets as venting spaces for collective antagonism. Accordingly, commandeering the streets of Cairo, the centre of the uprisings, amounted to an appropriation of city space for radical performance of crisis in three distinct ways. First, scenes of street protest publicized by media reports at the time popularized street violence (Al

DOI: 10.4324/9781003122098-6

Jazeera 2013; Lynch, Glasser, and Hounshell 2011; Shenker 2011a, 2011b). Images of the wounded, the dead, and heavily tear-gassed civilians were used to procure audience attention and empower the narrative of state repression at a moment of imminent political change. This emphasis on street violence illustrated a crisis that, for many Egyptians, had become a norm. Second, although it proved hazardous and temporary, such occupation restored the dignity of citizens by positioning them above political domination. Publicizing crisis disarmed the state of its legitimate power, affording new practices of unrestrained citizenship to manifest. Third, the streets' occupation transformed city residents from passive captives to active urban activists. For the protestors "already caught in the nets of state 'discipline'" (de Certeau 1984, xiv–xv), the street actions amounted to tactical manoeuvres and explicit performances of revolt as an emancipatory gesture. The streets were, in this respect, a stage where the everyday crisis was re-inscribed and operationalized as a meaningful response to the political fabric of the nation. Within the dynamics of 2011–2013 crisis in Cairo, a contestation over spatial hegemony overflowed to a discourse on citizenship rights, exposing the implicit tactics of state-led repression as a disciplining mechanism and citizens' resistance as a revolt discourse.

Within this conflict, a new conversation of the meaning of civilian actions has become possible: that the "nation itself is no longer a successful arbiter of citizenship" (Holston and Appadurai 1996, 188). The street protests in Cairo transfigured the ideas of public space, positing the everyday as a system of signs about the social, the political, and the economical (Abaza 2014, 165). Michael Goldman (2015, 63) describes these protests as hints that a "new world was possible, one free of autocratic rule, free of austerity and speculation, free of multiple forms of oppression." Per this thinking, these protests constituted a dialectic of the longstanding conflict that achieved historical priority under the aegis of the Arab Spring. Cairo's streets, contested by the state and citizens, present an incoherent city that is experienced as a clash of ideals originating with revision of the political mandate and proceeding with a determination to use this leverage to improvise terms of citizenship. Beyond the signification of social dualism and mechanisms of control, these street actions configure an inherent crisis not exclusively expressible in spatial practices but embodied by the protestors themselves. The ideals of citizenship pursued through the uprisings sought not much benefit from occupying the streets but rather relied on the streets to reify the residents' capacity to mobilize, to act, to refuse the norm, and to embrace radical manoeuvres as a necessary option to negotiate with the state. This article pursues a less-acknowledged form of crisis beyond the global spectacle of street violence: the subtle expression of repression and undercover resistance in Cairo cinema, which orientates us to a longstanding norm of citizens' suppression. I call this process of normalized suppression "crisis citizenship."

Crisis citizenship refers to the public performances by the state or citizens aimed at determining the models of citizenship that are possible. Through the use of violence as disciplining mechanisms, the state seeks to enforce its authority by subjugating citizens. In response, citizen mobilization as a tactic of resistance may undercut the legitimacy of state power. In this framework of discipline and revolt, an overt tug-of-war between citizen and state arises. Crisis citizenship is thus a useful conceptual tool for theorizing "the urban nature of these expressions of people's power and social change" (Goldman 2015, 63). In the case of Cairo, Egypt, these expressions are configured by contiguous historical conflicts of state formation: first, when the government fought external interference and second, in subsequent internal power wrangles between government structures (Gordon 1992). Two versions of revolution – a political revolution from foreigners and a revolution to provide social justice for all – comprise the contiguity of this history (Ginat 1997, 124; Nasser 1955, 26–27).

Thus, in the pre-2011 context, the tussle between state and citizens hints at a historical debt to *al-thawra al-muzdawija* or a revolution that can guarantee progressive socio-economic inclusivity. In Cairo, this debt is located within the revolutionary youth group, *shabāb al-thawra*, under the aegis of the April 6th Youth Movement (Rennick 2015). It is precisely this perception of a debt owed by the state to its citizens that led to the more recent uprisings, strictly graspable as a tussle over social (and economic) hegemony. Certainly, from the citizens' point of view, the 2011 and 2013 ouster of Egypt's presidents sought improved citizenship by the incumbent presidential officeholder, rather than a strict change of political power. Through the notion of crisis citizenship, we can critique the concessions sought through street protests not as efforts to correct incipient political crisis but as expressions of a need to recognize the coming-of-age of latent forms of activism to rebuild citizenship welfare. Khaled Marie's *Asal Aswad* (*Molasses*) (2010) is exemplary of this new paradigm.

Asal Aswad

Asal Aswad is an Egyptian film produced by Company of the United Brothers Cinema. Its director, Khaled Marie, is an acclaimed Cairo-based filmmaker whose works reflect remarkable creative flair. His most notable films include *Taymour and Shafika* (2007), *Aasef ala el-iz'ag* (2008), *Bolbol Hayran* (2010), *Tamantashar Yom* (2011), and *Laaf Wa Dawaraan* (2016). He directed *Asal Aswad* not long before the street protests in Tunisia that broke a political taboo and which set the stage for the 25 January 2011 Egyptian uprising. In *Asal Aswad*, he tells the story of Masry Sayyed Al-Arabi (Ahmed Helmy), a 30-year-old photographer returning to Cairo from the United States where he has lived for 20 years. Although Masry has both Egyptian and American citizenship, on this trip he has travelled on his Egyptian passport (leaving his American passport) and thus is strictly identifying himself as an Egyptian.

Shaimaa Saied (2019) describes *Asal Aswad* "as a poignant reminder of our current times," while Nadia el Magd (2010) praises the "film's ability to relate to everyday Egyptians," thereby positioning the narrative within the realism of urban citizenship in Cairo. However, within the 2010 context of a politically volatile Egypt, just before the onset of the 2011 and 2013 political protests, this film seems to presage on a simmering crisis rather than merely reflecting 2011's urban Egypt. Although *Asal Aswad* has subsequently been read as a film about cultural hybridity (Al-Hassani 2016) as well as a migrant-returnee comedy (Reesh 2015), in view of the 2011 events that it pre-empts, its full meaning cannot be confined to the events of 2010 Cairo. It is more speculative of a future crisis underway than it is an actuality narrative. Likewise, the bulk of its meaning is not found in the themes of poverty, corruption, oppression, and general urban chaos that it conveys but in its use of plot and mundane urban experiences as subtexts of a serious national imaginary.

Dina Mansour (2012, 3) has argued that "film relies on precise social, political and cultural contexts to reflect and mirror contemporary moral values, norms and attitudes while dramatizing existing societal problems." We may not dismiss film narratives as merely artistic, but we can also see them as dynamic languages of, among others, political and social imaginations of a particular place and time. To the extent that we consider Egyptian films, which are part of the Arab world cinema, as "texts ... produced by history" (Khatib 2006, 2), we can read such film narratives as important insights into national consciousness, whether overtly expressed or subtly coded. Historically, Egyptian films have not only kept abreast of the country's political times but also have become potential sites for conveying such underlying national imaginaries. Pre-revolution Egyptian films are seen as "representations in the service of ideological positions," and thus as "enabling an uprising that was itself a result of many factors" (Tabishat 2012, 377). In her discussion of Khaled Youssef's *Heen Maysara (Waiting for Better Times)* (2007), Nouran Al-Anwar El-Hawary (2014, 57) argues that due to the "extremity of social disparities between the classes in Egypt, people have long feared the uprising of those people from *ashwayet*, commonly named '*thawret il giyaa*'" (the revolution of the hungry). She cites the film's depictions of inequalities as potentially insightful.

Accordingly, this chapter critiques *Asal Aswad*'s use of the experiential city as a subtext of "vacant nationalism" (Abu-Samra 2010). Focusing on the subtle rendering of governance and citizenship problems in pre-revolution Cairo, through the protagonist's mundane actions, it explores how contemporary urban Arab cinema communicates sensitive political imaginaries, such as simmering revolt. Masry's heartbreaking experiences in Cairo give the film an anticipatory tone while retaining a narrative restraint typical of the films of the period. Through these overtures, this film immerses the viewer in a *Kefaya* (enough) pre-revolution Cairo, that is, a tipping point for civilian anger fed-up with state repression, frustration, and humiliation, and hence using the city as a stage for national change (Khamis and Vaughn 2015, 300). In Cairo, Kefaya (also known as the Egyptian Movement for Change) became popular in 2004 as a protest movement targeting political issues. Its most notable focus at the time was to prevent President Hosni Mubarak from running for a fifth term or transferring power to his son, Gamal Al Din Mohammed Hosni Ei Sayed Mubarak (Shorbagy 2007, 41). Outside Egypt,

Kefaya was seen as an indigenous reform movement whose singular focus on Mubarak's hereditary politics broke down Egyptians' "aversion to direct confrontation with the regime" (Oweidat et al. 2008, ix). Yet, while this movement relied on public mobilization and performances, the consciousness of revolt had for a long time been part of the city's private imaginary.

It is thus enriching to think of how *Asal Aswad* uses the character of an individual revolutionary youth as an embodiment of collective grievances. Alongside the story of everyday urban pressures in *Asal Aswad*'s Cairo, the "meaning of revolution – the specific vision of power, state-society relations, and the nature of radical change" is embodied by Masry, who personifies the "prefiguration (or prefigurative politics) and the manifestation in the present of future changes sought" (Rennick 2015, 34). In this sense, then, Masry's everyday actions, being neither overtly political nor forceful, sustain the Kefaya temperament of an individualized, even customized, response to overwhelming issues aptly resonating with the ideals of *shabāb al-thawra,* already cited. The citizenship issues raised in the film – corruption, bureaucracy, lack of essential services or poor government services, lack of freedom, poverty, and a general condition of despair – are potential registers of the shrewd oppression underway in the everyday. For this reason, the protagonist's street-level, non-violent actions encapsulate the citizenship manoeuvres underway in pre-2010 Cairo as emblematic, indeed pre-emptive of unvocalized revolution underway. The subtlety of his actions, however mundane, supports the film's function as a cautious resistance piece. To foreground this assertion, this chapter discusses the two forces that sustain dramatic tension in *Asal Aswad*: the state's combative force that pressures the protagonist and the protagonist's subtle force, which sensitizes the viewer to his crisis citizenship and motivates a mindset of revolt.

Revolt under construction

The city space, writes Robert Cowherd (2008, 275), is "both a manifestation of social dualisms and an instrument through which they are maintained and reproduced." He further adds that social dualism is founded on "social distinctions and relies on urban space as a means of control" (Ibid., 276). Unmistakably, Cowherd's point is that urban space is neither idle nor a tabula rasa, but is endowed, through usage, proximities, and assemblage of infrastructure, with the capacity to convey human dichotomies and the mechanisms that sustain such distinctions. This idea is cognate with Henri Léfebvre's (1991) and Michel de Certeau's (1984) well-known theories of urban space. Specifically, social dualism intersects with Léfebvre's (1991, 165) concept of spatial appropriation – challenging the existence of solid, urban orderliness marked by a schism between different socio-economic tiers and instead transforming the streets to activism spaces that can be seized by disenfranchised groups seeking the possibility of a more cohesive experience. Léfebvre frames this melee in his canon of the rights to the city. Equally, Cowherd's thoughts on space and control can be transposed with de Certeau's (1984, 30) philosophy that walking the city spaces constitutes "ways of operating" in the city or guidelines for using the city space. To exist in the city space is to be tasked to operate in a certain way, to uphold or probe its orderliness. Per these scholars, urban space has the possibility of characterizing the way urban citizens respond to social and political dualism. In practice, this is indeed intrinsic to most postcolonial African cities, including Cairo (Abu-Lughod 1965, 429). Here, spatial experiences are filtered through an infra-structural design reifying a system of inclusion and exclusion (Ghannam 2002) and labour laws that sustain oppressive tendencies among the urban poor (Chalcraft 2011). In response to these and other disenfranchising condi-tions, Cairo's population occasionally inverts the spatial guidelines that sus-tain its schismatic orderliness and use radical protest tactics (Lubeck and Britts 2002, 335) like spatial occupation.

In Cairo, Tahrir Square protesters, unified by shared citizenship ideals, coalesced into a single symbol of radical protest. Here, "the reconfiguration of social and political relations within the autonomous space of the 'Republic of Tahrir' ... was collectively interpreted by the activists as the revolutionary ideal" (Rennick 2015, 167). Given this occupation of the streets and other public spaces throughout what has since been termed as the Arab Spring, Mona Abaza (2014, 166) argues that the "Arab revolutions have triggered a stimulating debate about inventing new and real public spaces that merge with virtual imaginary, public spaces which occur through collective perfor-mances and actions." Here, one infers the significance of the streets and other public spaces in marking not only the evolutionary configuration of city spaces into political texts of imagined freedom but also the occupation of the streets as a signifier of progress towards imaginary political victory. This overt crisis was captured on mobile phone cameras, broadcast on global media, recorded in films, and seen in the images of the wounded, the faces of defiance, the shouts of determination, and in the violent clash between opposed groups and between the state and civilians.

Conversely, revolt under construction – how the revolution was embodied in the everyday long before the 2011 street uprisings – offers a different con-ceptual lens to account for covert forms of repression and resistance. The disenfranchised citizen is the infrastructure of revolt under construction, embodying an implicit conflict between state repression tactics on one hand and undercover citizens' resistance on the other. If, as Dirk Kruijt and Kees Koonings (2009, 25) point out, "violence transforms urban politics and limits the scope for empowerment of the urban excluded," then violence as a metatext for combative negotiation of hegemonic power between state and citizens within cities can only reveal a partial picture of resistance in Cairo. Revolt under construction emerges from the need to illustrate latent expres-sions of revolt in pre-2011 Cairo. In this period, overt protests were minimal and ineffective (Khamis and Vaughn 2015), yet, as Katia Janjoura reports on *Connected in Cairo* (2012), Egyptian filmmakers were aware of "a revolu-tion-in-waiting, simmering under the surface of Egyptian society." A more revealing fact is that pre-2011, to express whatever grievances the citizens may have, filmmakers faced heavy government censorship designed to pre-serve "public norms, religion and culture" (Mansour 2012, 1) and to control the "representation of existing social and cultural realities" (Ibid., 2). Youssef Chahine, one of Egypt's iconic filmmakers, at one point had multiple films – *The Sparrow* (1973), *The Emigrant* (1994), and *Chaos* (2007) – banned by state censorship (Bradley 2008, 169–170) because they were considered potentially provoking.

Mohammed Tabishat (2012, 377) notes that while from as early as 1992 Egyptian cinema was already telling stories about "anxiety, mistrust, and fear" between socio-economic groups, this transpired within an awareness of the censorship norm requiring filmmakers to be constantly alert to the pre-cautions expected of the medium. Yet, despite these restrictions, filmmakers such as Tamer El Said, who directed *In the Last Days of the City* (2016), continued to narrate what Arab News (2017) termed as the "impotent rage" in Cairo. On their website they report that a month and half after El Said finished filming, "the same Cairo streets exploded into protest." In a rather limiting turn of events, the website further reports, the film could not be screened in Cairo. The reason, per the director's words quoted by Arab News (2017), is that "There is just no place at the moment for an alternative voice." It is useful to note that this reporting is dated 26 June 2017 – about 6 years after Mubarak was ousted and 4 years since Morsi was also ejected. The consequences of this strict censorship on Cairo's representability are twofold. On one hand, the hesitant representation of Cairo's everyday nuances in film is a testament to the fact that dissenting voices have no ready means. On the other hand, such concessions over verisimilitude within the films compelling filmmakers to use covert creative symbols to pass their messages across sug-gest the coerciveness of censorship mechanisms. From these discussions, it appears problematic, even misleading, to discuss cinematic narratives of cri-sis citizenship exclusively through images of street violence or the media's activist accounts. Despite the fact that social media journalism enabled activ-ist mobilization and popularized the protest (Khamis and Vaughn 2015) and that films popularized the revolt as an obligatory response to repressive regimes, it is pre-revolution films that played an even bolder role. In these films, the affinity for the verisimilitude of repression, as seen in the street performances of overt resistance found in post-2011 films, is largely absent. Instead, these acts are replaced with visual political crypts. Revolt under con-struction thus offers an alternative avenue by which to read crisis citizenship beyond tropes of violence.

The right to Cairo as latent repression

Post-2011, Egyptian filmmakers have used Cairo streets as staging sites for combative performances of state repression and civilian resistance. Among the all-inclusive is the short film collection *18 Days* (2011), featuring *Retention* (dir. Sherif Arafa, 2011), *God's Creation* (dir. Kamla Abou Zikri, 2011), *19-19* (dir. Marwan Hamed, 2011), *When the Flood Hits You* (dir. Mohamed Ali,

2011), *Curfew* (dir. Sherif El-Bendary, 2011), *Revolution Cookies* (dir. Khaled Marie, 2011), *Tahrir 2/2* (dir. Mariam Abou Ouf, 2011), *Window* (dir. Ahmad Abdallah, 2011), *Interior/Exterior* (dir. Yousry Nasrallah, 2011), and *Ashraf Seberto* (dir. Ahmad Alaa, 2011). These films liberally render the citizen-level experiences of the 2011 uprising, illuminating the emancipatory models of citizenship deployed beneath the aura of violence and chaos: unity, freedom, conviviality, and empowerment. Other notable films of this time include *The Square* (dir. Jehane Noujaim, 2013), *Tahrir 2011: The Good, the Bad, and the Politician* (dir. Ayten Amin, 2011), *Uprising: The Birth of the Egyptian Revolution* (dir. Fredrik Stanton, 2012), *From Queens to Cairo* (dir. Sherif Sadek, 2012), *In Tahrir Square: 18 Days of Egypt's Unfinished Revolution* (dir. Jon Alpert & Matthew O'Neill, 2012), *Crop* (dir. Johanna Domke & Marouan Omara, 2013), *Waves (III)* (dir. Ibrahim El-Batout, Corrado Sassi, 2012), and *In the Last Days of the City* (dir. Tamer el Said, 2016). These films, like the short film collection mentioned above, used the images of tear-gassed and wounded civilians, some fatally, exhibiting laudable boldness in publicizing crisis citizenship. The media showed a similar inclination.

Dominic Nahr, working for TIME and Magnum in Motion, created a montage of still photographs of the early days of the uprising against Hosni Mubarak, overlaying it with sounds of explosions, voices of the agitated crowds, and other audio elements (Time Photo Department 2011). The result was an enhanced verisimilitude of violence and emphasis on the overt practices of citizen oppression in Cairo. In an episode of *People and Power* focusing on street protests in Cairo, Al Jazeera (2013) used images of the military attacking protestors, including the brutal final showdown in which bulldozers ploughed through civilians and their makeshift structures to emphasize overt violence as authenticating state brutality. Certainly, these post-2011 films and journalistic accounts of the revolution by global media offer reportage-style narratives that do not shy from the verisimilitude of a vast violence tapestry. The films use the images of street violence to authenticate repression and achieve verisimilitude of the struggles involved in appropriating Cairo streets by its civilians (Gordon 1992, 26–27). The problem, however, is not to recount the facileness of street protests, as this has, to a larger extent, been extensively covered. Rather, the challenge is to explicate the ways in which these violent street protests were merely the climax of a crisis that existed long before erupting in the streets. Thus, while parlaying a publicly popular image of the Egyptian revolution, such combat cannot account for other forms of repression and resistance that were in place well before the 2011 street uprisings. *Asal Aswad* embodies this approach.

This film starts with the final approach segment of Masry's flight to Cairo. In a series of interior inflight shots, we see a flight attendant distributing immigration arrival forms. Whereas the commercial airplane as a form of transport may just construe the immigratory aspect of the scene, the camera is positioned and framed to emphasize Masry (identifying with his Egyptian passport) and his seatmate (Mohamed Shahin, identifying with his American passport), thus visually pinpointing the pair as the most important elements in the frame. Pairing two Egyptian characters who possess dual citizenship yet identify with different citizenships moments before landing, and positioning this juxtaposition in the opening montage of the film, gives urgency to the office of immigration. Through the issuance of entry forms in the plane, the film establishes a connection between the characters and Cairo's airport immigration desk. This filmic moment foreshadows latent repression in three ways.

First, it can be located within Manuel Castells's (1977, 111) concepts of localizing inequalities, the "processes of articulation between the urban units (citizen categories) and the system of producing social representations and practices (state mechanisms)," and the struggles to institutionalize citizens' rights. Identifying himself as an Egyptian citizen, Masry gives the state authority to determine his experiences of citizenship, by contrast, identifying as an American, as his seatmate does, precludes this authority. This juxtaposition elicits a paradox of citizenship status versus the state's latent repression. Through his seatmate's advice that Masry should have brought his American passport, and Masry's question, "Why should I bring it and be treated as a foreigner in my country?", the film is cuing the viewer on the possible interference of citizenship rights in Cairo. The history of citizen–state tussle is thus implicit in the conclusion that Cairo is a "meeting ground and battleground for two opposed worlds" (Triulizi 1996, 81): the imagined world and the encountered world. It is a city where circumventing crisis citizenship may require existing as an underground citizen by adopting the identity of a foreigner.

Second, the characters' imminent encounter with Egypt's airport immigration desk, with its administrative and regulatory power to decide who enters Cairo, may be read within Léfebvre's (1991, 154) theory of the right to the city. Against the backdrop of his utopic proclamation of loyalty to the Egyptian state, Masry's experiences at the immigration desk reveal the state's coercive forces as modes of repression. While Masry's seatmate eases past the immigration desk, Masry waits for a long period for clearance. Such a denial of rights to an Egyptian citizen and his decision to reinstate his American citizenship, concretely exemplify the film's use of symbolic narration to convey pre-revolution crisis citizenship. The tendency of the protagonist to seize his dual citizenship and exist "somewhere in the middle" (Khatib 2006, 173) as an Egyptian-in-the-making and an American-in-the-unmaking, influence the way one would imagine this film as a political narrative. With the protagonist's ever-imminent risk of being branded a non-Egyptian, the film sustains his characterization as "perhaps" representative of the Egyptian psyche. In fact, the protagonist exists mostly as a stateless character, so that his actions, though clearly based on his initial expectations of extensive rights as an Egyptian citizen, are toned down to appear as incidences of an American character struggling to cope in Cairo. Henceforth, it becomes clear that while Masry's experiences are not atypical, they also do not constitute a political voice per se, but rather humanize the widespread civilian crisis for all to see. It is through this creative manoeuvre that the film underwrites the aesthetics of self-conscious cinema being used to "tackle the political dilemmas and social problems of contemporary Egypt" (Schochat 1983, 22), while also sufficiently throttling the political discourse to evade censorship.

Third, Masry's prolonged waiting deploys "time wasting" as a form of withholding his rights, hence exhibiting a repressing, disciplining tool. This act registers the state's capacity to intimidate, control, punish, and render powerless its citizens. It is a hegemonic activity: the state mobilizes the power of its agencies to downgrade Masry's right to Cairo below even the "simple visiting right" (Léfebvre 1991, 158) enjoyed by his "American" seatmate. Thus, for the duration that he uses his Egyptian passport, Masry does not really "enter" Cairo – as marked by the reunion with his family's neighbours. That is, he exists as a visitor: staying in hotels, sightseeing in the pyramids on a hired vehicle and a stallion – these aspects mark him as a tourist, a visitor. Subsequent scenes such as when he is arrested and jailed for taking pictures in Cairo, feed into this structure of visitation rights and oppression. Unable to even pay his own bail, he relies on Radi (Lotfy Labib), the taxi driver, to bribe the police. Consequently, when we see him offering a bribe to skip the long queue when applying for an Egyptian national identity card, we can identify his complicity as an exposé of the pervasive oppression among state authorities. Such moments index the state's use of institutional capacities to overrule its own mandate over citizenship rights. Although Masry's private experiences contradict, for instance, the Charter of National Action (*mithaq al-'amal al-watani*), which promised better prospects post 1952 (Ginat 1997), they hardly draw attention to the political discourse within the film. Instead, his experiences passively construe an imploding regime besieging its own citizens, thus providing a baseline for shared injustices (Clarke 2011). Yet, while the state's treatment of Masry may illustrate tyrannical tendencies, the rhetoric of state authority sustained throughout the film can also be read as a creative attempt to expose discreet private resistance.

Indignity and undercover resistance

Shimelis Bonsa Gulema (2013, 191) has rationalized that "the city or its fragments like the street and the crowd are constructed as a place of resistance and inversion ... all entangled in the dialectic of 'interpenetration.'" Certainly, this statement invites scrutiny into the everyday street encounters between state and citizens as concurrently constituting interpenetration. In Cairo, it is not easy to use the term revolution without invoking its connotation of political spectres. In fact, the term was almost censored from government discourse, appearing in 1953 to signify the "gulf between new and former rulers" (Gordon 1992, 192). In response, the revolution has existed as a conflict beyond the violent display of state authority or civilian protests in the streets. It is embodied in covert state and civilian practices and experiences, surfacing sporadically in, among other forms, the normalized indignities of everyday urban life. As such, *Asal Aswad* addresses these pertinent everyday urban issues, such as urban mobility, housing, and citizenship.

On urban mobility, the film shows the indignity of travelling in Cairo's public bus, an experience that many Cairo residents would identify with. To travel in the state-operated city bus (*otobis*), writes Farha Ghannam (2002, 1):

> [y]ou need to know how to jump when it slows down as it nears your station. You also have to know how to jump into the bus before it speeds up and joins the flow of traffic. But above all, you need to learn how to fit your body among the masses on board while paying close attention to your belongings. In addition, you need to acquire the skill not only of quickly grabbing any vacant seat but also of sharing it with young children and older people.

This mode of transport, which Ghannam contrasts with the comfort of travelling by taxi, metro, or microbus, has come to typify the struggles and constraints of Cairo's residents when navigating their city. Deciding to travel in otobis, in many cases lured by the lower ticket price, is choosing to travel in indignity. The scene of Masry travelling in otobis in *Asal Aswad* is easily recognizable as an instantiation of Ghannam's theory of urban mobility in Cairo. It invites the viewer to perceive the pervasiveness of daily indignities in Cairo's public transport. The interior shots of Masry pressed between crooks and subsequent exterior shots of him jumping from the moving otobis invoke his undignified mobility as a means to publicize crisis citizenship in a tight-lipped historical moment. In later sequences, we see Masry, his neighbour, and children riding an overcrowded scooter motorbike to school – again paralleling the indignities of the otobis. Consequently, his experiences of indignity in Cairo – the exorbitant room charges, the car rental company subjecting him to excessive leasing conditions, a good stallion being exchanged for a bad one at the pyramids – all because he identifies himself as an Egyptian, amounts to a codeword for publicizing Cairo's ineffectuality.

On public housing, one notices in *Asal Aswad* that Masry and his neighbours live in an alley, considered in Egyptian cinema as an allegory of the national citizenship crisis (Shafik 2012, 94). Beyond identifying Masry as an underclass citizen, his private moments potentially exploit crisis citizenship for a cathartic result, namely, disarming the state's veil of political legitimacy (Triulizi 1996, 81). *Asal Aswad* seizes the struggles of the alley as a microcosm of Cairo's housing crisis, metaphorically illustrating an ineffectual state.

Finally, on citizenship, *Asal Aswad* uses two scenes to intimate private and public denouncement of Masry's Egyptian citizenship. Privately, Masry denounces his Egyptian citizenship early in the film, when we see him on a hotel balcony discarding his Egyptian passport after receiving his American one. It is a simple expression of conquest: the protagonist stands on the balcony of the hotel one morning, in a white nightgown, and slings his Egyptian passport to the streets below using an elastic band tied on the rails to mimic a catapult. Typically, a catapult is a device that throws objects as projectiles, either to hit designated targets or to vanish. In this action, the catapult amplifies the notion of resistance, partly extending the perimeter of the action beyond the balcony to implicate the streets below, the dumping site for his Egyptian passport (and hence the detestations associated with it). This private–public trajectory connotes the state's diminishing importance as the protagonist dissociates from the citizenship associated with state control, and marks Masry, who is seen in the foreground from the vacuous city, as dissociated from the state authority (denoted by the highly policed city). To show Cairo as a defocused background lacking any specific point of focus or visual intelligibility gestures to the way the experiences of crisis citizenship indexed by the city are no longer central to the protagonist's world. This dramatic action is shown as a response to the indignities he has endured so far. Catapulting his Egyptian passport intimates a zealous rejection of oppression, and indicates Masry's defiance of these experiences. It also conveys his exasperation with his Egyptian identity and the state, of which the passport is emblematic, an attempt to salvage the citizenship rights denied to him. Yet, this very crucial gesture is presented as a private, nonchalant action where the viewer is the only spectator. In this case, the action of discarding his passport with a catapult serves as a metaphor of Masry's covert disassociation from the Egyptian state, while the private nature of the action renders it as an undercover action to which the state is not privy.

Thus, to speak of such a moment as instantiating undercover resistance springs from the conceivable significance of other private actions, whether the indignities of the otobis or in the alley family space or the private frustration with citizenship as reactions constitutive of the Kefaya (enough) activist temperament. Without overemphasizing the coding of indignity, one may argue that its private nature and the corrective manoeuvres that it coerces among the city's underclass – as embodied by Masry's action of throwing away his Egyptian passport – wholly constitute a reaction to the Egyptian state. If these instances are seen as evoking the underlying tension between the protagonist and the state rather than as sporadic coincidences, they become "privileged sites for considering the current renegotiations of citizenship" (Holston and Appadurai 1996, 188–189). As such, they expose the ineffectuality of pro-citizen narratives trumpeted at various historical junctures.

Masry's publicly denouncement of his Egyptian citizenship occurs shortly after this private episode. That is, after a minor accident in which the car of a wealthy man bumps the rear of Radi's taxi in which Masry is travelling to look for his family friends, a state police officer lets the offender leave and instead punishes Radi. Infuriated, Masry confronts the officer, who refers him to his seniors nearby, whom Masry also accosts, declaring that he is now an American citizen and thus beyond their reproach. Nearby, a procession of protestors carrying banners with anti-American slogans can be seen. It is this group of protestors that assaults Masry, taking his American passport and abandoning him in the street. After the mob disperses, Masry is lying on the ground, his shoes and jacket scattered around him, a banner emblazoned with the words "DOWN WITH AMERICA" nearby.

This scene renders a public conflict between the state and its citizens. That the police choose to procure mobsters to "discipline" Masry rather than offer justice amounts to a denial of justice. It suppresses the citizenship ideals which Masry seeks: freedom and justice. These ideals are not encouraged and it is noteworthy and a provocative gesture that the mobsters start off as protestors (a symbol of revolt) but get influenced by the state (represented by the police force) to become mobsters against Masry (who personifies the entitled American). This progression exemplifies the uncertain, even unstable, conception of discipline and revolt by the state, hence positing uncertainty as a potentially repressive tactic. If Masry, identifying as an Egyptian or an American, cannot access his perceived justice, what ideas of the state does this convey? Bearing in mind the exceptional interest in the idea of citizenship and governance cultivated within the film, Masry's public vexation with his citizenship experiences is suggestive of the universality of oppression. Like his private protest, this public confrontation with the state is a personal way to signal an underlying psyche of exasperation with the established norms of civilian–state relations. Questioning this norm amounts to exposing it as a form of tyranny. When read alongside his private denunciation of the state (as seen by his throwing away the passport), such a dramatic moment suggests the value of undercover manoeuvres as a more practical approach to resisting the state than taking public action.

Hence the close-up shot in the final sequence of the film, which emphasizes a wedding ring that Masry wears, merits attention. Why would he, after exposing the state's ineffectuality and registering his discontent throughout the film, wear a ring adorned with the colours of Egypt's national flag? Why is the engagement ring, a symbolic prop, introduced at the film's end and not at the beginning? If we read the ring as a souvenir, a way of memorializing Masry's feelings about Cairo, we start to see it as a meaningful symbol of the way Masry feels towards Cairo. Whereas the colours of the ring identify him as a citizen "intimately" connected to his country, the notion of political nuptiality to which the ring is indexical merits more attention. If we read Masry's frustrations throughout the film as provocations for him to abandon his quest for better experiences in Egypt, then his choice to declare his strong relationship with the state by getting engaged to "her" expresses commendable resilience. Instead of quitting his rather activist life in Cairo and returning to America, he chooses to stay. Hence, the film's ending resonates with his characterization as Masry Sayyed Al-Arabi, the Egyptian Arab, suggesting his personal commitment to continue his subtle resistance rather than abandon the cause. Accordingly, the film's ending suggests a form of defiance, a wait-and-see scenario that prefigures the 2011 public protests.

Drawing on Henri Léfebvre's and Michel de Certeau's spatial theories, this chapter has critiqued Khaled Marie's *Asal Aswad* as a film that foreshadows crisis citizenship in pre-2011 Cairo. It has discussed the film's use of visual metaphors of the state's latent repression and citizens' undercover resistances in the form of disidentification with the Egyptian state, as illustrating a latent exasperation. The overall argument is that pre-2011 films invested in creative ways to express exasperation with state-imposed experiences of citizenship while showing remarkable restraint. The first part of the chapter discussed the idea of crisis citizenship, positioning *Asal Aswad* as a film narrative exemplifying this concept. The chapter's second half critiqued the institutional checkpoint represented by the airport immigration desk as a prototype of state control. Further, the subsection critiques Masry's experiences of indignity in the areas of urban mobility, housing, and citizenship as an exposé, and hence a criticism of the government's shortcomings. It is argued that his decision to remain in Cairo at the end of the film illustrates a personal commitment to the latent revolt seething underneath everyday life in Cairo. The conclusion is that many of Masry's actions are markers of expiation, potentially revelatory of the underpinning civilian responses to political tensions in pre-revolution Cairo. This film works through such political and social visual metaphors to underpin undercover practices and means of countering the state's authoritarianism.

Bibliography

Abaza, Mona. 2014. "Post January Revolution Cairo: Urban Wars and the Reshaping of Public Space." *Theory, Culture & Society* 31 (7/8): 163–183.

Abu-Lughod, Jane. 1965. "Tale of Two Cities: The Origins of Modern Cairo." *Comparative Studies in Society and History* 7: 429–457.

Abu-Samra, Haisam. 2010. "Assal Eswed: A Fish out of the USA." *Cairo 360*. July 6, 2010. https://www.cairo360.com/article/film/assal-eswed-a-fish-out-of-the-usa/.

Al-Hassani, Huda Abdullah Abdulateef. 2016. "Cultural Differences in Hybridity: A Study of Asal Eswed (Black Honey) and New York Movie." *Semantic Scholar*. Global Proceedings Repository. https://pdfs.semanticscholar.org/7d5b/61b7d9eb1a ac57b1a838cefbaa5b79de60fe.pdf.

Al Jazeera. 2013. "Egypt: Revolution Revisited: How the Dramatic Events in the Summer of 2013 Have Affected Those Who Brought about Egypt's 2011 Revolution." *Al Jazeera People and Power*, September 16, 2013. https://www.aljazeera.com/programmes/peopleandpower/2013/09/2013913144518879459.html.

Arab News. 2017. "Epic Struggle of Film About Cairo on Eve of Revolution." June 26. http://www.arabnews.com/node/1120526/offbeat.

Bradley, John R. 2008. *Inside Egypt: The Land of the Pharaohs on the Brink of a Revolution*. New York: Palgrave Macmillan.

Castells, Manuel. 1977. *The Urban Question: A Marxist Approach*. London: Edward Arnold.

Chalcraft, John. 2011. "Labour Protest and Hegemony in Egypt and the Arabian Peninsula." In *Social Movements in the Global South: Dispossession, Development and Resistance*, edited by Sara C. Motta and Alf Gunvald Nilsen, 35–58. Hampshire: Palgrave Macmillan.

Clarke, Killian. 2011. "Saying 'Enough': Authoritarianism and Egypt's Kefaya Movement." *Mobilization: An International Quarterly* 16 (4) (December): 397–416.

Connected in Cairo. 2012. *Documentary Films for Teaching About the Egyptian Revolution*. October 16, 2012. https://connectedincairo.com/2012/10/16/documentary-films-on-the-egyptian-revolution/.

Cowherd, Robert. 2008. "The Heterotopian Divide in Jakarta Constructing Discourse, Constructing Space." In *Heterotopia and the City: Public Space in a Postcivil Society*, edited by Michiel Dehaene and Lieven De Cauter, 275–286. London and New York: Routledge.

de Certeau, Michel. 1984. *The Practice of Everyday Life*. Translated by S. Rendall. Berkeley: University of California Press.

el Magd, Nadia Abou. 2010. "Popular Film Holds Mirror to Egyptian Society, for Better and Worse." *The National*, June 18, 2010. https://www.thenational.ae/world/africa/popular-fim-holds-mirror-to-egyptian-society-for-better-and-worse-1.532369.

El-Hawary, Nouran Al-Anwar. 2014. *The Graffiti of Mohamed Mahmoud and the Politics of Transition in Egypt: The Transformation of Space, Sociality and Identities*. Masters Thesis, Cairo: American University in Cairo.

Ghannam, Farha. 2002. *Remaking the Modern: Space, Relocation, and the Politics of Identity in a Global Cairo*. Berkeley and Los Angeles: University of California Press Berkeley.

Ginat, Rami. 1997. *Egypt's Incomplete Revolution: Lufti al-Khuli and Nasser's Socialism in the 1960s*. London: Frank Cass.

Goldman, Michael. 2015. "Development and the City." In *Cities of the Global South Reader*, edited by Faranak Miraftab and Neema Kudva, 54–65. London and New York: Routledge.

Gordon, Joel. 1992. *Nasser's Blessed Movement: Egypt's Free Officers and the July Revolution*. New York and Oxford: Oxford University Press.

Gulema, Shimelis Bonsa. 2013. "City as Nation: Imagining and Practicing Addis Ababa as a Modern and National Space." *Northeast African Studies* 13 (1): 167–213.

Holston, James, and Arjun Appadurai. 1996. "Cities and Citizenship." *Public Culture* 8: 187–204.

Khamis, Sahar, and Katherine Vaughn. 2015. "Cyberactivism and Citizen Mobilization in the Streets of Cairo." In *Cities of the Global South Reader*, edited by Faranak Miraftab and Neema Kudva, 300–303. London and New York: Routledge.

Khatib, Lina. 2006. *Filming the Modern Middle East: Politics in the Cinemas of Hollywood and the Arab World*. London and New York: I.B. Tauris.

Kruijt, Dirk, and Kees Koonings. 2009. "The Rise of Megacities and The Urbanization of Informality, Exclusion and Violence." In *Megacities: The Politics of Urban Exclusion and Violence in the Global South*, 8–26. London and New York: Zed Books.

Léfebvre, Henri. 1991. *The Production of Space*. Translated by Donald Nicholson-Smith. Oxford: Blackwell.

Lubeck, Paul M., and Bryana Britts. 2002. "Muslim Civil Society in Urban Public Spaces: Globalization, Discursive Shifts, and Social Movements." In *Understanding the City Contemporary and Future Perspectives*, edited by John Eade and Christopher Mele, 305–336. Oxford: Blackwell Publishers.

Lynch, Marc, Susan B. Glasser, and Blake Hounshell. 2011. *Revolution in the Arab World: Tunisia, Egypt, and the Unmaking of an Era*. Washington, DC: Slate Group.

Mansour, Dina. 2012. "Egyptian Film Censorship: Safeguarding Society, Upholding Taboos." *Alphaville: Journal of Film and Screen Media* (4): 1–16.

Nasser, Gamal Abdel. 1955. *Egypt's Liberation*. Washington, DC: Public Affairs Press.

Oweidat, Nadia, Cheryl Benard, Dale Stahl, Walid Kildani, Edward O'Connell, and Audra K. Grant. 2008. *The Kefaya Movement: A Case Study of a Grassroots Reform Initiative*. Santa Monica: RAND Corporation.

Reesh, Mohammad Ali Abo. 2015. *Analysis for Selected Comedy Films in Egyptian Cinema*. Masters Thesis, Gazimağusa, North Cyprus: Eastern Mediterranean University.

Rennick, Sarah Anne. 2015. *The Practice of Politics and Revolution: Egypt's Revolutionary Youth Social Movement.* Doctoral Thesis, Lund: Lund University.

Saied, Shaimaa. 2019. "Movie – Assal Eswed – 2010." *El Cinema*, October 31, 2019. https://elcinema.com/en/work/1313173/content.

Schochat, Ella. 1983. "Egypt: Cinema and Revolution." *Critical Arts* 2 (4): 22–32.

Shafik, Viola. 2012. "From Alley to Shanty Town: Representing the Nation Through Cairo's Changing City-Scape." In *Afropolis: City, Media, Art*, edited by Kerstin Pinther, Föster Larissa, and Christian Hanussek, 92–97. Auckland Park: Jacana Media.

Shenker, Jack. 2011a. "Cairo Protesters in Violent Clashes with Police: Egyptian Protesters Call for End to Hosni Mubarak's Rule and Hail 'First Day of Revolution'." *The Guardian,* January 26, 2011. https://www.theguardian.com/world/2011/jan/25/egypt-protests-mubarak.

Shenker, Jack. 2011b. "In Tahrir Square of Cairo Freedom Party Begins: Jubilant Egyptians Celebrate Hosni Mubarak's Resignation." *The Guardian*, February 11, 2011. https://www.theguardian.com/world/2011/feb/11/tahrir-square-cairo-freedom-party.

Shorbagy, Manar. 2007. "Understanding Kefaya: The New Politics in Egypt." *Arab Studies Quarterly* 29 (1) (Winter): 39–60.

Tabishat, Mohammed. 2012. "Society in Cinema: Anticipating the Revolution in Egyptian Fiction and Movies." *Social Research* 79 (2) Egypt in Transition: 377–396.

Time Photo Department. 2011. "The Uprising: Reflections on the Egyptian Revolution by Dominic Nahr." *TIME,* April 15, 2011. http://time.com/3775980/the-uprising-reflections-on-the-egyptian-revolution-by-dominic-nahr/.

Triulizi, Allessandro. 1996. "African Cities, Historical Memory and Street Buzz." In *The Post-Colonial Question: Common Skies, Divided Horizons*, edited by Iain Chambers and Lidia Curti, 78–91. London and New York: Routledge.

Filmography

18 Days. Dir. Ahmad Abdalla, Maryam Abu-Of, Kamlah Abu-Zikri, Muhammad Ali, Sharif Arafah, Sharif El-Bindari, Ahmad Ala El-Dib, Marwan Hamed, Khaled Marie, Yusri Nasrullah, Egypt, 2011.

19-19. Dir. Marwan Hamed, Egypt, 2011.

Aasef ala el-iz'ag. Dir. Khaled Marie, Egypt, 2008.

Asal Aswad (Molasses). Dir. Khaled Marie, Egypt, 2010.

Ashraf Seberto. Dir. Ahmad Alaa, Egypt, 2011.

Bolbol Hayran. Dir. Khaled Marie, Egypt, 2010.

Chaos. Dir. Youssef Chahine, Egypt, 2007.

Crop. Dir. Johanna Domke & Marouan Omara, Egypt, 2013.

Curfew. Dir. Sherif El-Bendary, Egypt, 2011.

From Queens to Cairo. Dir. Sherif Sadek, Egypt, 2012.

God's Creation. Dir. Kamla Abou Zikri, Egypt, 2011.

Heen Maysara (Waiting for Better Times). Dir. Khaled Youssef, Egypt, 2007.

In Tahrir Square: 18 Days of Egypt's Unfinished Revolution. Dir. Jon Alpert & Matthew O'Neill, Egypt, 2012.

In the Last Days of the City. Dir. Tamer El Said, Egypt, 2016.

Interior/Exterior. Dir. Yousry Nasrallah, Egypt, 2011.

Laaf Wa Dawaraan. Dir. Khaled Marie, Egypt, 2016.

People and Power. Dir. Al Jazeera, Qatar, 2013.

Retention. Dir. Sherif Arafa, Egypt, 2011.
Revolution Cookies. Dir. Khaled Marie, Egypt, 2011.
Tahrir 2/2. Dir. Mariam Abou Ouf, Egypt, 2011.
Tahrir 2011: The Good, the Bad, and the Politician. Dir. Ayten Amin, Egypt, 2011.
Tamantashar Yom. Dir. Khaled Marie, Egypt, 2011.
Taymour and Shafika. Dir. Khaled Marie, Egypt, 2007.
The Emigrant. Dir. Youssef Chahine, Egypt, 1994.
The Sparrow. Dir. Youssef Chahine, Egypt, 1973.
The Square: The People Demand the Downfall of the Regime. Dir. Jehane Noujaim,
    U.S. and Egypt, 2013.
Uprising: The Birth of the Egyptian Revolution. Dir. Fredrik Stanton, Egypt, 2012.
Waves (III). Dir. Ibrahim El-Batout's Ibrahim El-Batout, Corrado Sassi, Egypt,
    2012.
When the Flood Hits You. Dir. Mohamed Ali, Egypt, 2011.
Window. Dir. Ahmad Abdallah, Egypt, 2011.

7 Outlier urbanism

Inside Luanda's postwar cantonments

The modes of inhabiting city space in postcolonial Luanda is historically eclectic. Even without extreme and obvious spatial rules dictating modes of urban tenancy, covert demarcations still exist in postwar Luanda (Mututa 2019). Such delineations ensue from the ever-present existential barriers to accessing the infrastructure necessary for favourable urban survival. This imaginary permeates many postmillennial films about the city. Postwar Luanda materializes in José Augusto Octávio Gamboa dos Passos' (Zézé Gamboa) *O Herói (The Hero)* (2004) as a city fortified by intrinsic yet unseen barriers. The survival of the residents of postcolonial Luanda appears, at a glance, to be constituted along rigid, politicized social and economic structures. Thus, understanding popular cultural imaginaries of Luanda as a restrictive city also implicates the underhand workings of the postcolonial welfare state and the implications of its soft power (Watson 2002, 49). Without delving into the political economies of such relations between the state and postcolonial urbanism in Angola, the question of citizenship rights – or the intersection of the politics of postwar urbanism and political manoeuvres – is the subject of this chapter, which explores the representations of the realism of fortification in postwar Luanda in Gamboa's *O Herói*. It argues that the protagonist's confinement in wretched urban life – evident in his futile pursuit of healthcare, job opportunities, and better living conditions – is emblematic of the project of exclusive post-colonial urbanism in Luanda. Using cantonment and outlier urbanism as a conceptual lens, the chapter critiques *O Herói's* representation of Luanda's urban form and the protagonist's experiences in various urban sites, as offering useful perspectives on Luanda's postwar urbanism. That film characters' oscillation towards a better life is juxtaposed with a sense of despair, it is argued, portrays an elaborate scheme of alienation within the city. The theorization that follows is twofold. First, that urban characters coping with inescapable incapacitation in their pursuit of postcolonial inclusion symbolize a verisimilitude of postwar cantonment; and, second, that such urban restriction produces peculiar urban practices, which I theorize as "outlier urbanism." Both aspects are paradigmatic of Luanda's postwar urbanism.

DOI: 10.4324/9781003122098-7

Luanda in cinema

For most of Angola's colonial history, especially from the 1930s, the country's film production has been dominated by Portuguese filmmakers. The natives were, to a greater extent, consumers of foreign Portuguese films screened in a vast network of cinemas. Some of these cinemas, albeit in various states of disrepair, can still be found in Luanda today. That said, local filmmakers like Mariano Bartolomeu, despite having only intermittent involvement in Angola's film production, have made remarkable impressions on the country's film industry. Bartolomeu's renowned films include U*n lugar limpio y bien iluminado (A Clean and Well-Lit Place)* (1991), *Quem Faz Correr Quim (Who makes Quim Run?)* (1991), and *O Sol Ainda Brilha (The Sun Still Shines)* (1995). Angola's other three most prominent homegrown films were all released shortly after the end of the country's three decades of civil war in 2002. Among them is Orlando Fortunato de Oliveira's *Comboio da Canhoca (Train of Canhoca)* (1989), shot towards the tail-end of the war, but released in 2004. Narrating the story of 59 Angolans and the Portuguese secret service holding them hostage, it focuses on the stress that ensues between these erstwhile state captives when their boxcar is detached from the train and is abandoned for 3 days in the railway yards of Canhoca.

The plot of Fortunato's film is built around the desecration of the native through confinement in oppressive space. The boxcar in which these natives are now restrained can be read as a metaphor for Angola's postcolonial urban condition. Their restriction in an unpleasant space and the condition of their abandonment on a railyard is emblematic of postcolonial urban helplessness. A similar representation is to be found in the director's earlier films. *Memória de um Dia* (Memory of a Day) (1982) tells the story of the Portuguese massacre in Bengo Icolo in 1960, while *Batepá* (2010) is about the 1953 Sao Tome e Principe killings of native *forros* (creoles) by the Portuguese. In these films, Fortunato seems to use cinematic elements to reflect on the realism of war and the uncertainty inherent in its possible aftermath. His keen interest in narratives of subjugated natives evident in these films brands him as a historical filmmaker, while his visual narratives of Angola's postwar difficulties underscore the constraints of fictionalizing difficult histories such as Angola's. Thus, although his narratives smooth out the vigour of the actual experiences of turmoil they recount, the films show the unmistakable nuances of Angola's difficult postcolonial transition. Such creative conflict between fictionalizing history and historicizing fiction through the techniques of the cinema medium forms the basis for critiquing the representations of postwar urban relations in subsequent films set in Luanda.

This nexus between history and fiction is to be seen in the films of Maria João Ganga, another prominent Angolan filmmaker whose film *Na Cidade Vazia (The Hollow City)* (2004) in 2004 won both the Special Jury Prize at the Paris Film Festival, and the most promising filmmaker at the Cape Town World Cinema Festival. It uses images of Luanda's cityscape to recall and recast subtexts of postwar crises through the urban form. This film tells the story of N'dala (Roldan Pinto João), a young boy evacuated from the warfront of Bié in Huambo to Luanda. Upon landing in Luanda, he escapes from his custodians and wanders through the city, leading the viewer through diverse spaces and experiences. The dilapidated shacks, the lunatic character, the dark lighting and shadows, the police patrolling the city, the hawkers selling their merchandise in the open streets at night – all index various facets of Luanda's chaotic urbanism. Noteworthy is the coexistence between the film's past (its narrative of wartime Luanda) and its present (its production in the postwar times) and the reflections that this comingling enables. One of these reflections is how postcolonial Luanda is characterized by excepted citizenship. Here, the old fisherman (Custodio Francisco) who lives in a grass structure at the ocean-front is emblematic of how Luanda's residents are trapped in a perpetual state of indeterminacy and speculation where nothing is guaranteed. He restores the "struggle of the postcolonial citizens to cope with their situation of exception, seen here as the mainstream formula of inhabiting Luanda" (Mututa 2019, 97). By characterizing the city as such, the film explores the gap between postcolonial urban ambitions and the ever-elusive postwar urban prosperity.

This representation of persistent distress, particularly as a counter-narrative archiving Luanda's new postcolonial order (Harrow 2013, 203), supports the analysis of Luanda's crisis urbanism in Gamboa's *O Herói*. Financed by the Angolan government, this film won the World Cinema Dramatic Jury Prize of 2005 at Sundance Film Festival. Set in the streets of Luanda, the film tells the story of Vitório (Oumar Makena Diop), a former military sergeant who lost one leg after stepping on a ground explosive during the war, he has now come to Luanda looking for a means to survive. From the film's beginning to end, Vitório is stranded in unfavourable experiences: lack of a job, exclusion from urban communities, and lack of basic amenities such as accommodation, sanitation, or proper healthcare. The ideals of heroism implicit in the title of the film, contrasted with the protagonist's despicable urban realities, open up a space for dialogue that Vitório is a hero not because of his wartime credentials (which he discards) but because of his capacity for resilience that motivates him to survive in such a city as Luanda. The film's visualization of Luanda through such a protagonist forms part of Angola's post-civil war narratives, which reinstate the historical memory of urban precarity as very much a present if differently modelled urban reality.

O Herói accomplishes this by refreshing the memory of Luanda's recurrent oppression. The film has been read as a "window into post-civil war society and the social decay and familial disruption produced by three decades of conflict," and at the same time appraised as a "realistic if hopeful look at how individuals accommodate the dislocations of war and rebuild their lives amid the rubble of contemporary Angola" (Parrott 2014, 233). This reading is further supported by Zézé Gamboa's interview, cited in Rudi Rebelo (2014), where he describes the postwar city:

Estes 45 anos de guerra foram muito pesados, deixaram cicatrizes que vão demorar muito a sarar. A minha responsabilidade de cidadão e cineasta é a de chamar a atenção da sociedade civil e das autoridades para lhes dizer que isto não pode voltar a acontecer. Acho que para exorcizar a lembrança da guerra O Herói é um filme importante para todos os angolanos. Sendo uma ficção, mostra o estado problemático do país neste momento.

(These 45 years of war have been very heavy, leaving scars that will take a long time to heal. My responsibility as a citizen and filmmaker is to draw the attention of civil society and the authorities to tell them that this cannot happen again. I think to exorcise the memory of the war *The Hero* is an important film for all Angolans. Being a fiction, it shows the problematic state of the country at the moment).

If by "exorcising memory" Gamboa means to overlay the wartime images of misery with postwar symbols of hope drawn from Luanda's cityscape, then the film's historicizing of Luanda's postcolonial urban dysfunctionality extends the context of exorcising to become atemporal. Particularly, the struggles of the diverse characters in the film – all constituting Luanda's diegetic urban fabric – convey a peculiar imaginary where Luanda's postwar urbanism does not fulfil the expectations of the majority of its residents but instead creates a peculiar crisis. Beyond the crisis of urban poverty (Cain 2012, 15), urban inequalities (Gastrow 2017; Peel 2013), urban marginalization (Beinart, Delius, and Trapido 1986; Jones 2000), and urban segregation (Knox and McCarthy 2005, 22), a new crisis of inescapable incapacity to rise above undesirable urban conditions is now more noticeable among the city's majority. This is perceptible in the characterization of the motley crew of urban misfits populating Gamboa's film.

There is Vitório, an ex-military officer who faces great difficulties integrating into civilian life. Surviving as a cripple, he lacks both a social community and economic means. Through him, the film evokes the paradox of Angola's postwar urbanism: that many have emerged from a debilitating period to the realism of insurmountable incapacity. He is characterized as a peripatetic urban misfit. There is his lover, Judite (Maria Bárbara Simões), who works as a prostitute and stays in a dingy space. Throughout the film, Judite's experiences reflect the sadness with which survivors of the war inhabit Luanda as a subterfuge; a place where, despite offering hope for a better life, one does not heal, but draws comfort from congregating amongst others with similar problems. Then there is Manu (Milton "Santo" Coelho), a young boy whose father went missing in the war, and who is being raised by his ageing grandmother. In the film, Manu gradually loses interest in school and joins his peers to form street gangs. Their most remarkable crime is a deal in which they gain possession of Vitório's prosthesis. These young boys, like N'dala in *Na Cidade Vazia*, signify the cyclic hopelessness in which generations of Luanda's residents inhabit the city. Not only was the past difficult but the present and the future it presages also appear equally bleak. These representations of a slice of Luanda's postwar urban community are significant when discussing cinematic representations of Luanda's postcolonial urbanism in *O Herói*. In pursuing this community, the film is not merely making a display of Luanda's postwar difficulties but articulating the place of human experience in conferring meaning to Luanda's postwar urbanism. In the discussions that follow, I theorize the inextricable confinement in hopeless exclusion as a form of what I term "urban cantonment," and the resulting urbanism characterized by insuperable susceptibility within poor urban quarters as "outlier urbanism."

Cantonment in *O Herói*

A term with military origins, "cantonment" refers to exclusive army installations, such as barracks or encampments, both temporary and permanent. I use it here to infer confinement in specific urban situations and experiences as well as the spaces where such experiences are embedded. In postwar Luanda, the term implies controlled access to services and opportunities on the basis of insider-outsider groupings. It also designates the history of assembling Luanda's residents in particular spaces, material experiences, and sociality. When former UNITA soldiers were demobilizing at the end of the civil war, many of those who had nowhere to go or were simply wary of joining their communities were segregated into cantonments. In November 2002, Human Rights Watch (2003) reported that "over 445,000 former UNITA soldiers and their families lived in forty-two camps located around the countryside." These quartering areas, inhabited by both former UNITA military officers and civilians, produced a peculiar urban form where certain populations were restricted to specific urban experiences. From this history, Luanda's modernization has produced selectively appropriated "new social 'enclaves', closed and guarded residential spaces" (Rodrigues 2009, 37) as physical markers of urban segregation. These also enforce "asymmetrical relations of power" (Keith and Pile 1993, 38), producing unequal urban possibilities. Cantonment in postwar Luanda, then, prefigures the social and economic assembling of Luanda's residents into neighbourhoods and communities, and the incapacity to overcome such circumstances in the absence of external intervention or political bestowal. Accordingly, cantonment is best understood as a circumstance of hierarchical rights where "most rights are available only to particular kinds of citizens and exercised as the privilege of particular social categories" (Holston 2007, 7). This situation has been described as an "uncomfortable truth ... (that) some enjoy full citizenship rights while others remain marginalized from them" (Martins 2017, 100). The task at hand is to theorize cantonment in Luanda as a practice that did not lapse with demobilization but rather metamorphosed into an urban condition that continues to shape the urban form and experiences therein.

Films about Luanda abstract this situation in a number of ways. Maria João Ganga's *Na Cidade Vazia* (2004) uses characterization to convey urban entrapment in unfavourable conditions. The film's most poignant characters include an aged fisherman (Custodio Francisco), Joka (Raúl Rosário) and his criminal friends, the prostitute Rosita (Júlia Botelho), the orphan Zè (Domingos Fernandes Fonseca), and N'dala, the former surviving on the benefaction of strangers, and the latter evacuated from Bié. An unemployed pilot also shows up at the end of the film. This characterization of Luanda's residents suggests the insurmountable boundaries that have come between the city's urban poor of all generations and their pursuit of success in the postwar city. Further, it is significant that the narrative ends not with their transformation but instead their deterioration: Joka and his friends are validated as criminals after breaking into the pilot's house, Rosita's drunkenness has increased; N'dala is shot dead, Zè has started drinking and smoking, and the fisherman's fate is altogether forgotten. While the significance of the *casebres* (a hovel or shack) and *musseques* (informal settlements) as visual symbols in this film have been well discussed as expressions of exceptionalism (Mututa 2019), the idea of being fixed in the condition of urban squalor, to which they gesture, is largely under-discussed. Thus, the everyday life of characters in this film may be read both as an expression of struggles for staking claim to a social, economic, and political territory, and as a way to challenge cantonment, which became a mainstay urban practice since the end of the civil war. It is for this reason that cantonment is used here as a useful metaphor to read Luanda's hierarchical modernization as producing a limiting urbanism. This is particularly useful in reading cantonment in *O Herói*.

O Herói uses ex-military officer Vitório's struggles and precarious life in Luanda as emblematic of his confinement in the city's postwar exclusion. Particularly, his inability to rise above marginalization throughout the film caricatures the persistent postwar exclusion of Luanda's residents from postcolonial opportunities. Vitório's most noticeable challenges include the loss of his prosthetic leg to criminals early in the film, the subsequent long wait to get a replacement for the prosthetic leg in the city public hospital, his attempts to get employment, and his inability to forge a real urban community where he can belong. Although these incidences are poised as Vitório's difficulties in reintegrating with civilian life, they reveal the barriers that exist in attaining a better gradation of rights. Above all, they ruminate on the barriers that the protagonist faces in postwar Luanda as a first step to interrogating the film's setting – including the poor musseque communities – as abstractions of Luanda's postwar crisis urbanism.

In Kenneth Harrow's (2013, 17) critique of the film, he notes that the "lives of *O Herói's* characters are the products of a process that cannot be dissociated from the global strategies of the great powers and from the global economics," thereby creating a basis to discuss the film's characterization as concurrently emblematic of Luanda's prevailing urban crisis and its historicity. Thus, when he avers that the characters embody the "heterogeneous" trash generated by the conflict" (Ibid.), he is juxtaposing the city's war history with its postwar conditions. He further pursues the idea that the film is a mimesis of postwar urban crisis by discussing the symbolism of Vitório's prosthesis as the

> Device for rehabilitating a body that has lost its value, having become trash, having been trashed; and as the reiteration or copy of the original contents of the archive, a revision to a missing or absent original entry that reappears at a later time
>
> (Ibid., 203)

He concludes that, as the primary symbol in the film, the prosthesis signifies an aporia bred of insufficiencies and dysfunctionality, in that even when Vitório learns to use it "his handicapped gait (remains) a sign that the supplement cannot fill the lack but only indicate its presence" (Ibid., 206). If Vitório personifies Luanda's postwar citizen, then, his condition of irrecoverable loss indexes the inescapable conditions of despair, and hence the aporia of Luanda's postwar urbanism.

This aporia is most evident in *O Herói's* characters' catastrophic failure to achieve better urban prospects, and its visual style – dominated by urban alleys, musseques, and other dilapidated spaces. At the start of the film, a panning high-angle camera mimicking a bird's eye view of Luanda reveals a vast landscape of musseques dotting an otherwise arid landscape with undulating hills. Clearly, this vast musseque landscape underscores the predominance of informal communities in postwar Luanda in the postcolonial times when the government was popularizing inclusivity. It was a remarkable coincidence (if a coincidence at all), that soon after the film won a Sundance Film Festival prize, Angola's second President Jose Eduardo dos Santos announced the building of one million modern homes for Luanda's residents in 2008 – a move that appeared to be a promise to abolish the city's division of citizenship experiences along material aesthetics. As argued in subsequent scholarship, the housing projects – starting with the first one constructed after the president's promise, *Nova Cidade de Kilamba* (Kilamba New City) – have come to nurture socio-spatial patterns that continue to promote unequal experiences of citizenship in the postcolonial city. The new housing projects did this by supplanting centre-periphery structures with emergent norms of social and economic capabilities as a formula of assigning urban socio-economic groupings. *O Herói's* opening montage, which shows the musseque landscape and halts just as the main city is coming to view, is a premonition of this unfruitful urban life in Luanda.

Cognizant of the aporic value of this initial sequence, it is arguable that Gamboa's film takes a second glance at the spatial images that have defied the post-independence political rhetoric of a prosperous and inclusive Luanda. It takes a jab not only at the seeming indifference to the boundaries that link the cidade and the musseque (the camera's continuous shot stops just as after we see the modern buildings, thus creating a sense of spatial contiguity despite material differences) but also establishes a visual context for interrogating Luanda's cityscape and its conglomeration of zones as a beacon of more urgent issues that the protagonist espouses in his peripatetic urban journey. The indexicality of vulnerability and inferiority produced by the "God's eye view" (Gomes and Abreu 2017) adopted by the camera to show the musseque communities as a wasteland, focalizes on the unequal urban form as the mainstream image of the cityscape. In this context, the presence of Vitório in this cityscape supplants the image of Luanda as a cohesive modern metropolitan with a self-portrait (a selfie if you may) of his exclusion from this glamour.

This representation of the city prompts two questions: how do we make sense of such post-2000 images of Luanda and the verisimilitude of dystopia that they suggest? And how do we interpret this inescapable misery when it no longer belies pursuit of freedom but expresses a norm among Luanda's residents? These are, of course, questions that touch on Luanda's urban social life, material inequality, and politics of space and which determine the relationships that are possible between different socio-economic enclaves. However, the filmic discourse on these issues – evident in the images of the vast musseque landscape and Vitório's initial appearance as a cripple – suggest the portability of marginalized identity (Gomes and Abreu 2017, 154) from war to postwar Luanda. On the literal level, this crippling pronounces debilitating urban life as a mainstream postwar reality. In retrospect, however, developing Vitório's character alongside extreme difficulties and his inability to egress this lowly economic enclave problematizes a larger urban crisis that can be discussed in two ways.

First, Vitório's socio-economic immobility bares the politicization of inside–outside political citizenship, requiring him to submit to the city's "necessary but demeaning 'progress'" (Bose 2008, 40). In Luanda, being categorized as *povo do governo* (pro-government) or *povo da UNITA* (pro-UNITA) assigns economic and hence social inclusion or exclusion (Pearce 2015, 15, cited by Martins 2017, 103). Vitório's homelessness and disability make him an ideal vehicle by which to highlight this interplay of exclusion and political subterfuge. The actions of selective political intervention that the film explores by linking Vitório's socio-economic mobility to political intervention suggest a nucleated political and economic linkage from which citizens are excluded by default. It is not enough that Vitório is a "former revolutionary, the liberation guerrilla fighter, … or the autochthone," he must back up the "maintenance of the present political status quo of the elites who control access to citizenship rights" (Martins 2017, 101) in order to access this inclusion. Characterizing Vitório as a poor, postwar urban dweller – rejected, jobless, homeless, and disabled – and restricting his socio-economic mobility to the benevolence of the Ministro do Interior (Orlando Sérgio) help us to see his urban citizenship as being crafted, contested, and concretized through covert political patronage. In other words, conferring socio-economic volatility upon Vitório is largely a narrative strategy to elicit urban marginality as a façade to exchange Luandan's bourgeoisie dreams of inclusivity with political submission to political elites (Martins 2017, 101). This symbiosis procures economic crisis and underpins the perpetuity of Vitório's urban misery as a postwar crisis out of which one cannot easily escape, if ever.

The second point is around Vitório's limits or how the condition that spells out his urban life "emerges situationally" (Gastrow 2017, 227). The situation here is exclusion as a form of political coercion. Vitório's material deficiency is inseparable from his political indebtedness. His ideals of a modern postwar Luanda are "only attainable through a promise that never fulfils itself" (Tomás 2012, 297), suggesting Luanda's social and economic categories as congealed networks. From *O Herói's* initial representation of the city's edge as a vulnerable frontier, to its ending with government-mediated socio-economic mobility, the film legitimizes political control, harnessing and manipulating marginality to strengthen the political oligarchy. Vitório's struggles to access medical services, employment, and social belonging, and that his attainment of socio-economic progress is reliant on "state economic apparatus" (Martins 2017, 111), provide a logic to interpret his crisis in postwar Luanda as a consequence of pervasive solipsistic manoeuvres. Vitório and the Ministro do Interior straddle a city "divided, while being interconnected by differential access to resources" (Gulema 2013, 183) – so that the transmutation of his urban experiences from dire lack to opulence, indexes an interstitial circumstance where politics and urbanism intersect.

The above discussions provide three perspectives useful in discussing Luanda's outlier urbanism. First, that the arbitrary boundaries that sustain urban differentiation between the elites and the poor simultaneously reflect underlying mechanisms of regulating access to resources, opportunities, and privileges. Thus, cantonment is not just an urban condition but an aperture into how a city's social and economic patterns concurrently constitute a system of classifying citizens into clusters of privileges. Second, that segmented citizenship experiences and trajectories also involve questions of rights, and hence of underlying political economies of urban governance. Cantonment, in this case, indicates the tussle to govern cities in particular ways and to serve specific political ends. Third, that enforcing unequal political, social, and economic participation facilitates different pathways of participation in national discourses. The layer that one occupies in the urban privilege circle also defines the limits to one's integration in the city's socio-economic structures. Cantonment is, by and large, about barricades that maintain urban clusters of inequality, and how unfavoured urban residents confront such fortifications. Incidentally, cantonment also concerns how such barricades nurture spatial urban practices resulting in a peculiar urbanism that is wholly aligned to neither colonial nor postcolonial imaginaries. The next subsection discusses outlier urbanism as a direct outcome of these practices of social and economic cantonment in postwar Luanda.

Outlier urbanism

Abdoumaliq Simone (2004, 408) describes urbanization as a "thickening of fields, an assemblage of increasingly heterogeneous elements into more complicated collectives," further asserting that the "accelerated, extended and intensified intersections of bodies, landscapes, objects and technologies defer calcification of institutional ensembles or fixed territories of belonging." The underlying concept here is that urbanization is foremost an aggregate of many diverse aspects – people, spaces, economies, and infrastructure – all collectively contributing to the urban form. This networking of urban elements is also the formula that makes it possible for different elements to interrelate and share the city as a common space. Yet, beyond this seamlessness, an urban assemblage is also a paradox that permits urban misfits to arise. Labelling the city quota where such misfits reside as "extensive areas where the mixtures are so intense that it is not easy to make definitive attributions" (Simone 2014, 322), and at the same time designating them as the missing people, Simone tasks urban assemblage to account for various forms of urban exclusion. The urban residents whom he assigns the domain of missing people – "nurses, teachers, factory and service workers, police, storekeepers, technicians, drivers, and those who adamantly refuse to be officially employed" (Ibid., 322) – comprise a partial list that changes from region to region, from city to city, and even from suburb to suburb within the same city. To be applicable in the specific context of Africa's postcolonial cities, the list can be revised to cater for widespread populations of the missing people. For instance, a category of urban tenants who are unable to find employment irrespective of how hard they try – and they are by far the majority in today's African cities – constitute a common category of the "urban missing" in Africa. These are misfits confined outside the reach of opportunities, social amenities, government services, state welfare, and other "fruits of postcolonialism" in urban Africa.

In her discussion of Angolan cities, Cristina Udelsmann Rodrigues (2009, 37) traces a comparable trend in Angolan cities. She talks of the replacement of colonial segregated urban form that followed "racial, economic and social stratification" with post-independence "socially and economically mixed areas in the cities," and later the emergence of new urban exclusion based on economic delineation. These changing urban practices, most manifest in the creation and repurposing of urban infrastructure to support new ideals of urbanism, incidentally express the paradox of, among other Angolan cities, Luanda's new urban forms and the underlying anticipations of urban sociality. In turn, this stratification problematizes Luanda's underlying crisis urbanism since independence and certainly since the end of the civil war in 2002. Lilly Peel (2013) describes this setup in postwar Luanda thus: "Across the road the new houses, now renamed Nova Vida (New Life), are like a dystopian 1950s suburbia – matching pastel colours and manicured front lawns surrounded by the shacks of the dispossessed." This reportage shows the musseque (symbol of urban exclusion) as usurped by a more universal symbol: the unattainable new life in the postcolonial city. New Life apartments signpost the desired life that remains, as of then, out of reach. It connotes a worsening alienation, urban displacement, and distress. The image of the old grandmother contemplating the new house across the street while remaining homeless provides a useful metaphor of the workings of urban enclaves in fostering economic, social, and political differentiation (Abbott 2004, 120). The resulting urbanism characterized by uneasy coexistence between residents of the musseques and modern urban enclaves communicates the limits of integrating the city as a whole. Instead, the resulting binary city of social-economic nuclei and adjacent communities opens up new prospects for theorizing "the outlier" as the contemporary figure of Luanda's postcolonial urbanism.

Arguably, then, Simone's theory of the urban missing, which addresses global south cities generally, when juxtaposed with Rodrigues' theorization of Luanda's urbanism and Peel's journalistic work addressing similar trends, is apt in decoding Luanda's postwar urbanism. This is particularly useful in appreciating *O Herói's* construal of Luanda's postwar urbanism as a seesaw between hopeful expectations countered by the realism of perpetual incapacity to intercept anticipated social and economic opportunities. To exist within this seesaw, the city's postwar residents – symbolized by Vitório – have to find innovative ways to live with pervasive poverty, low-key support networks, and impromptu manoeuvres. They are never at rest, instead, they continually move back and forth from one challenge to the other. These urban nomads constitute a particular kind of urbanism of perpetual confinement in unfavourable circumstances that are utterly restrictive.

The shot where we first see Vitório is set in a hospital corridor. The mise-en-scène is comprised of two queues of patients seated on benches placed against the two opposite walls. The camera is positioned at a low angle so that the framing accentuates the image of the two columns of patients and the space between them. Then, Vitório's feet emerge from beyond the foreground and we see the rest of his body as he limps along with the space in the middle, hoisting himself on a pair of crutches. The camera is static as he limps to the background where, after a jump cut, the camera switches to a normal angle tracking shot as he emerges at the back end of the hospital. Here, he confronts the doctors over his delayed prosthetic leg, securing an appointment only when he identifies himself as a former military sergeant. To use the composition of the hospital corridor full of patients as a visual cognate of postwar Luanda imagines the city's urbanism, indexed by this vacuous hospital space, as a debilitating place. It renders verisimilitude to the postwar narrative of the city's lingering suffering, which the film cultivates through mise-en-scène of the musseque material form, characterization, and narrative plot. To fully appreciate this montage, we are tasked to question the purpose of starting the film with long shots of the barren musseque landscape and then medium and close-up shots inside a hospital. Further, it is equally important to question the significance of juxtaposing the protagonist with such a visual map of the city.

These shots of the hospital corridor, coming after the initial aerial shots of the musseque city, catalogue the realism of postwar nostalgia, which the director clearly elicits through his characterization of the protagonist. The foreground–background axis of the corridor surrounded by waiting patients renders a vignette of the temporal stagnation useful in unpacking the iconic significance of marginal characters later in the film: the young boys who mobilize themselves into gangs to carry out petty theft, the mechanics and labourers who undertake strenuous work with little pay, the prostitutes, the queues of citizens in the streets waiting for their turn to get water from the communal tap, and the open-air market vendors. Through these characters, we see this film as a narrative populated by urban citizens who experience the city's realism of healthcare problems, joblessness, and lack of services and amenities not as disruptions in their postwar optimism but as an everyday urban norm in postwar Luanda.

O Herói's choice of an ex-military figure for a protagonist reflects the role of the native soldier, seen as a liberator, as central to envisioning the realism of the problematic aftermath of the war in Luanda. Vitório's character development (built around the idea of insurmountable stagnation) exposes the barriers that exist in attaining "full" citizenship rights in postwar Luanda. His pursuit of healthcare services interrogates how specific urban sites – in this case, the public hospital – have come to index a sense of placelessness, that is, the inability to find a space where (poor) characters can settle and feel welcome within the city. Vitório's character arc, ailed not so much by his physical disability but by the disabling circumstances of social and economic exclusion which he encounters in Luanda, emphasizes his unreserved exclusion from the city's economic and social progress. Through Vitório's character, it is hard not to notice the musseque as the space to which he has been assigned, and the hospital as emblematic of his irredeemable deficiency. This sequence thus alerts us to the broad ideas of urban Luanda with which the film is working, theorized here as outlier urbanism.

Outlier urbanism indexes unfavourable urban conditions and experiences and the resulting urban groupings into contiguous yet experientially segregated communities (Rodrigues 2009). It designates urban exclusion and urban tenancy detached from the mainstream urbanism of modernization and support infrastructure. In *O Herói's* Luanda, outlier urbanism is prevalent among poor musseque residents for whom Luanda's modern city remains out of bounds. Outlier urban survival exposes the iconoclasm of postwar urban inclusion. Further, the longing for anticipated urban benefits – such as job opportunities, access to social amenities like healthcare, and proper residential structures – is the identifiable marker of outlier urban existence. In postwar Luanda, outlier urbanism is a consequence of prolonged cantonment in precarious urban conditions. *O Herói* uses this fixation to characterize Luanda's postwar urbanism through persistent nostalgia.

Pam Cook (2005, 2) defines nostalgia as a "state of longing for something that is known to be irretrievable, but is sought anyway." It is "predicated on a dialectic between longing for something idealized that has been lost and an acknowledgement that this idealized something can never be retrieved in actuality and can only be accessed through images" (Ibid., 3). In the seminar presentation, "Urban Nostalgia: Colonial traces in the postcolonial city of Luanda," given at Wits Institute for Social and Economic Research (WISER), Antonio Tomas describes Luanda's urban nostalgia as a "sort of nostalgia for the future, or the future enshrined in urban forms of the past." In such a description, we see Luanda's urban form – its streets, structures, public spaces, and even inner rooms – as providing a visual vocabulary for the different dimensions of marginality and longing for a better life.

Connecting Tomas' and Cook's discussions of nostalgia and how these resonate with *O Herói's* narrative of Luanda's postwar urbanism, we find a basis to discuss nostalgia as a stylistic tool used in this film to support the representability of Luanda's outlier urbanism. By using nostalgia as a stylistic device, *O Herói* produces a "time-warp effect in order to suggest that historical change has not necessarily moved society forward" (Cook 2005, 10). Instead, it has produced peculiar forms of urban segregation whereby characters cut off from the benefits expected of the main city "did not so much seek to integrate themselves into an overarching framework of efficacy and normalcy as much as they attempted to put together a functional co-presence of different kinds of initiatives and orientations to the city" (Simone 2014, 327). That Vitório and other characters who comprise Luanda's outlier community "constantly lives under specific threats and incompletion" (Simone 2016, 136) does not identify them as degenerate or de facto members of Luanda's mainstream urbanism but as registers of fresh perspectives on the African city (Pieterse 2011). Among these perspectives is the notion of outlier urban existence in Luanda. In *O Herói,* this is personified by the itinerant characters of Vitório and, to a comparable extent, Manu, who, despite inhabiting the city's ghetto, still harbour hopes for better prospects.

In this film, Vitório functions as an inventory of a widespread sense of perpetual loss in the postwar city. Harrow (2013, 209) reads Vitório's prosthesis – at the centre of his transformation from a military figure to a civilian, and also his tool of urban mobility – as an "object that supplements a loss, that fills a lacuna, that turns loss into recovery, that makes whole, and so on," further adding that it is also "an object of exchange that passes from one person to another, and in so doing brings them together into a relationship." It is similarly arguable that his pursuit of completion is simultaneously a pursuit of inclusion into a city in which he is a stranger. Thus, the prosthesis symbolizes his claim for inclusion into Luanda's socio-economic communities. Its movement among different poor urban communities traces the longing for mobility, thus metaphorizing a widespread desire among Luanda's poor residents to evacuate their current state of affairs, which are not different from wartime. Through his usage of the city (as a cripple looking for completion), Vitório characterizes both Luanda's barriers to better urban prospects and the metaphorical use of waiting as a marker of outlier urbanism.

Alongside the outlier musseque community which he represents, Vitório conveys the ongoing conversation between Luanda's past (of war trauma) and its future (premised on better, inclusive, progressive life). Luanda's postwar exclusion, which he personifies, extends beyond the literal pursuit of longing for material inclusion, to the ineffectuality of breaking through the barriers of social and economic mobility. On this basis, *O Herói's* narrative can be read as an "optimistic response to a specific place and time" (Parrott 2014, 234). If the specific place is Luanda and the time is the postwar period, then the optimism alluded to here is the protagonist's persistence in braving a city inimical to his existence. Vitório's life in the streets of postwar Luanda triggers the idea of a time-warped postwar Luanda of severe social-economic stagnation for its musseque communities. Vitório's existence relies on provisional families, as embodied by Judite, who houses him for a while, and Manu, the young boy whose father disappeared in the war and now is under the care of his grandmother. These affiliations, however, do not evacuate the characters from Luanda's postwar stresses. Contrarily, they popularize the identity of a lone city character as a befitting vignette of postwar Luanda and his prolonged confinement in unsatisfactory urban experiences as a pervasive postwar experience.

These characters render the timelessness of Luanda's prolonged and wide-spread confinement in squalor and the accompanying nostalgia for better urban prospects since Angola's independence. From 1975, when President António Agostinho Neto took over Angola's leadership from Portuguese colonial authorities, and throughout the years of subsequent civil war, many Angolans who relocated to Luanda were seeking better prospects beyond the ongoing war that was ravaging many parts of the country. Yet, these prospects remained elusive even after President José Eduardo dos Santos took over the country's leadership in 1979, leading the country through the civil war to its end in 2002. With the majority of Luanda's population that migrated during the war largely experiencing squalor, occupation of musseques is emblematic of this prolonged marginalization and the nostalgia accompanying these enduring social and economic boundaries. Rodrigues (2009, 52) notes that Luanda's "[u]rban space is now subject to a type of appropriation based on social differentiation, on economic and social criteria, which (re)create competing claims on space." Given that houses in Luanda symbolize "financial stability, social and political incorporation of the person who is building" (Gastrow 2017, 231), the musseque characters that populate Gamboa's film, and the shanty houses they occupy, signify their exclusion from mainstream economic benefits and their unstable claim to Luanda's postwar socio-economic benefits.

In conclusion, I draw attention to Vitório's last-minute economic progress. The montage is preceded by a radio studio scene where he is being interviewed alongside Ministro do Interior. The shots of his praise speech are juxtaposed with shots of other members of the musseque community listening to him on the radio. Soon after, we see him driving with Manu along a dusty street. Their journey has no destination. This ending prefigures the vanity of his bestowment of car and house, which is repeatedly contrasted with the rest of musseque residents still trapped in urban squalor. This contrast powerfully comments on the ephemeral, and indeed the illusory, nature of Vitório's economic bestowment. The ending thus amplifies Luanda's poor residents as comprising the lowly but majority urban community cantoned in various forms of marginality.

Bibliography

Abbott, Carl. 2004. "Urbanism and Environment in Portland's Sense of Place." *Yearbook of the Association of Pacific Coast Geographers* 66: 120–127.

Beinart, William, Peter Delius, and Stanley Trapido. 1986. *Putting a Plough to the Ground: Accumulation and Dispossession in Rural South Africa, 1850–1930*. Johannesburg: Ravan.

Bose, Brinda. 2008. "Modernity, Globality, Sexuality, and the City: A Reading of Indian Cinema." *The Global South* 2 (1) (Spring): 35–58.

Cain, Allan. 2012. "Angola: Participatory Mapping of Urban Poverty." *Development Workshop*, July 30. http://urban-africa-china.angonet.org/sites/default/files/website_files/allan_cain_-_participatory_mapping_of_urban_poverty.pdf.

Cook, Pam. 2005. *Screening the Past Memory and Nostalgia in Cinema*. London and New York: Routledge.

Gastrow, Claudia. 2017. "Cement Citizens: Housing, Demolition and Political Belonging in Luanda, Angola." *Citizenship Studies* 21 (2): 224–239.

Gomes, Catarina Antunes, and Cesaltina Abreu. 2017. "As Time Goes by or How Far Till Banjul: African Citizenship Aspirations." *Citizenship Studies* 21 (2): 151–166.

Gulema, Shimelis Bonsa. 2013. "City as Nation: Imagining and Practicing Addis Ababa as a Modern and National Space." *Northeast African Studies* 13 (1): 167–213.

Harrow, Kenneth W. 2013. *Trash: African Cinema from Below*. Bloomington and Indianapolis: Indiana University Press.

Holston, James. 2007. *Insurgent Citizenship*. Princeton: Princeton University Press.

Human Rights Watch. 2003. *Child Soldiers in Angola Following the Conflict*. https://www.hrw.org/reports/2003/angola0403/Angola0403-04.htm#P328_62526.

Jones, Peris Sean. 2000. "The Basic Assumptions as Regards the Nature and Requirements of a Capital City: Identity, Modernization and Urban Form at Mafikeng's Margins." *International Journal of Urban and Regional Research* 24 (1): 25–53.

Keith, Michael, and Steve Pile. 1993. *Place and the Politics of Identity*. London and New York: Routledge.

Knox, Paul L., and Linda M. McCarthy. 2005. *Urbanization: An Introduction to Urban Geography, 2nd Edition*. Hoboken, NJ: Pearson Prentice Hall.

Martins, Vasco. 2017. "Politics of Power and Hierarchies of Citizenship in Angola." *Citizenship Studies* 21 (1): 100–115.

Mututa, Addamms Songe. 2019. "The Casebre on the Sand: Reflections on Luanda's Excepted Citizenship Through the Cinematography of Maria João Ganga's Na Cidade Vazia (2004)." *Journal of African Cinemas* 11 (3): 95–111.

Parrott, R. Joseph. 2014. "Reviewed Work: *The Hero* by Zézé Gamboa." *African Studies Review* 57 (1) (April): 233–234.

Pearce, Justin. 2015. *Political Identity and Conflict in Central Angola: 1975–2002*. New York: Cambridge University Press.

Peel, Lilly. 2013. "Angola's Poor People Hit Hard by Urbanisation Crackdown in Luanda." *The Guardian*, May 10, 2013. https://www.theguardian.com/global-development/poverty-matters/2013/may/10/angola-urbanisation-crackdown-luanda.

Pieterse, Edgar. 2011. "Grasping the Unknowable: Coming to Grips with African Urbanisms." *Social Dynamics: A Journal of African Studies* 37 (1): 5–23.

Rebelo, Rudi. 2014. *«O Herói» de Zézé Gamboa, espelho político-social de uma Angola pós-guerra*, November 3. https://nossaavenida.wordpress.com/2014/11/03/o-heroi-de-zeze-gamboa-espelho-politico-social-de-uma-angola-pos-guerra/.

Rodrigues, Cristina Udelsmann. 2009. "Angolan Cities: Urban Resegregation?" In *African Cities: Competing Claims on Urban Spaces*, edited by Francesca Locatelli and Paul Nugent, 37–54. Leiden and Boston: Brill.

Simone, AbdouMaliq. 2004. *For the City Yet to Come: Changing African Life in Four Cities*. Durham and London: Duke University Press.

Simone, AbdouMaliq. 2014. "The Missing People: Reflections on an Urban Majority in Cities of the South." In *The Routledge Handbook on Cities of the Global South*, edited by Susan Parnell and Sophie Oldfield, 322–336. London and New York: Routledge.

Simone, AbdouMaliq. 2016. "The Uninhabitable?: In between Collapsed Yet Still Rigid Distinctions." *Cultural Politics* 12 (2): 135–154.

Tomás, António Andrade. 2012. "Refracted Governmentality: Space, Politics and Social Structure in Contemporary Luanda." PhD Thesis, USA. PhD Thesis, Graduate School of Arts and Sciences, Columbia University.

Watson, Sophie. 2002. "The Public City." In *Understanding the City: Contemporary and Future Perspectives*, edited by John Eade and Christopher Mele, 49–65. Oxford: Blackwell Publishers.

Filmography

Batepá. Dir. Orlando Fortunato de Oliveira, Angola, 2010.
Comboio da Canhoca (Train of Canhoca). Dir. Orlando Fortunato de Oliveira, Angola, 1989.
Memória de um Dia. Dir. Orlando Fortunato de Oliveira, Angola, 1982.
Na Cidade Vazia. Dir. Maria João Ganga, Angola, 2004.
O Herói (The Hero). Dir. Zézé Gamboa, Angola, 2004.
O Sol Ainda Brilha (*The Sun Still Shines*) Dir. Mariano Bartolomeu, Angola, 1995.
Quem Faz Correr Quim (*Who makes Quim Run?*) Dir. Mariano Bartolomeu, Angola, 1991.
Un lugar limpio y bien iluminado (*A Clean and Well Lit Place*). Dir. Mariano Bartolomeu, Angola, 1991.

8 Crisis urbanism and the future of African cities

Since the turn of the millennium, the study of metropolises has shifted from the global modernity expressed by European and American megacities to alternative frameworks of modernity (Roy 2009, 828). The concept of modernity, central to 21st century theorization of cities, has undergone significant revisions to allow for a more decentralized theory that acknowledges native modernity and recognizes the potential of theoretical dislocation from mainstream western perspectives. On this basis, Roy (2009) argues for an urban theory based on the analysis of the global urban. Such an approach to contemporary urban theorization, particularly in the so-called global south, is not only welcome (Pieterse 2014), it has become a common practice in recent times (see for instance, Appadurai 1996; Beall, Crankshaw, and Parnell 2002; Bradbury 2012; Comaroff and Comaroff 2014; De Boeck 2011; Edjabe and Pieterse 2010; Fredericks 2018; Fujita 2013; Gastrow 2018; Gilroy 1993; Harvey 2003; Madani-Pour 1995; Mbembe and Nuttall 2007; Murray 2008; Myers 2011; Oldfield 2014; Parnell 2014; Pieterse 2008; Simone 2004a). This historical effort has motivated this book, which theorizes urban crisis as a localized, persistent form of urbanism. The book has demonstrated through the discussions of six African cities – Johannesburg, Kinshasa, Nairobi, Monrovia, Cairo, and Luanda – that crisis urbanism usefully accounts for native forms of urban life in Africa, which are responding to local postcolonial situations. Crisis urbanism also enables a cross-disciplinary conversation between urban studies and cultural (film) studies, which, in the face of increasing film productions in postcolonial African cities, is beneficial. The book is anchored in Mbembe and Nuttall's (2004, 348) concept of Africa as a "sign in modern formations of knowledge." To discover the meaning of African cities requires us to invest in human stories, which open up a productive analysis of the essence of lived urban life and its imaginary. This is useful because it will enable us to understand postcolonial urban problems through emergent conversations about those problems. Broadly, then, the book has theorized postcolonial cities, the most rapidly developing forms of Africa's modernity (United Nations Population Fund 2007), as the most viable sources and archives of Africa's urban imaginaries.

The implication of the book's focus on postmillennial African films – a vastly increasing urban archive in Africa – as a dominant cultural signifier of emergent urban ways of life merits a brief discussion. With this vastly

DOI: 10.4324/9781003122098-8

increasing urban archive in Africa in mind, the book encourages a contemporary theorization of African cities that acknowledges and encompasses the creative ways in which film narratives engage with serious postcolonial urban issues such as identity, economy, power, violence, governance, and fortification. At their symbolic level, the films discussed here plausibly engage with sensitive national questions around inappropriate policies, and position the question of urban crises as one touching on emergent postcolonial urbanisms. This replay of urban conflict designates urban citizenship, characterized by crises and adaptations to such crises, as the new common sense of Africa's postcolonial urbanism.

The book in contemporary context

Current theories on Africa's urbanism tend to focus on the apposition between "those who take an apocalyptic view and those who display an irrepressible optimism about the possibility of solving the myriad problems that beset such cities" (Pieterse 2008, 1). The former is exemplified by the work of Kees Koonings and Dirk Kruijt (2009) and the latter by Garth Andrew Myers (2011). Other existing theories on postcolonial urban Africa tend to be thematic, as evidenced by labels such as "roguery" (Pieterse and Simone 2013), "edginess" (Charlesworth 2005; Kruger 2013), "disposability" (Myers 2005), "disorderliness" (Murray 2008), "slummy" (Davis 2006), "revanchism" (Smith 1996), "transitioning" (Mututa 2019), and "shadowiness" (Neuwirth 2005). While noting that such references and labels are useful tools to condense debates on various urban eventualities, I suggest that they do not explicate how crisis – so central to existence in any African city – motivates unusual and unpredictable practices beneath the systematic city form, rendering such practices as a peculiar form of emergent urbanism.

Without identifying with either the optimist or pessimist strand of postcolonial urban theories, this book adopted a more suggestive approach made possible by two notions central to its completion: conceptualization and argumentation. On conceptualization, it has theorized urban life through the perspective of literary discourses, meaning it eluded the strictness of, say, anthropological or sociological data-based studies of cities, which yield statistical information, but are not perceptive of the sensibilities evoked by certain kinds of urban life. On argumentation, the book draws heavily on urban narratives as "truer" apertures into how Africa's ever-increasing urban tenants confront their cites day by day, and on the kinds of urban practices that emerge from this confrontation. To the extent that urban films locate their narratives within the cities' crises, it is argued that they comprise urban cultural dimensions (Nas, de Groot, and Schut 2011, 7). That is, they mirror various urban messages passing through the urban mundane practices (Lyotard 1984, 15), and in turn are instructive of emerging Africa's postcolonial urbanism. It is from this relationship between Africa's postmillennial urban films and urban imaginaries that crisis urbanism – a kind of urban life and practice that is responsive to existing crisis – has been proposed and used as a theory. Although crisis urbanism is the sole theory that has guided the reading of the six African cities discussed here, it has produced a more "situated" reading of the peculiarities of postcolonial urban Africa. The benefit of this is twofold.

First, it has dislodged the existing north-south theories that rely on mostly economic and political binaries, especially in respect to postcolonial cities. The symbol of slum – the most glaring form of economic deprivation and social demarcation and the benchmark for current hemispheric classification – does not necessarily exclude the pockets of economic growth within the same spaces. David Drakakis-Smith (2000, 3) discusses a case of Vietnam's fragmented urban experiences, where some "relax with a cold Pepsi or Heineken while they watch live satellite soccer from the new Chinese superleague." Yet, hardly "a few kilometers away," others "still take as long to walk to market as their great grandparents did, and are forced to pull their children out of school because they cannot afford the US 50 cents monthly fee." This lack of a geo-taggable economic frontier signals the limits of applying existing north-south theories, termed as "endemic to relational thinking on cities, [which] ... has influenced how world cities have come to be conceptualized" (Taylor 2004, 7). Certainly, the global south is beyond the grasp of economic theorization.

Second, and this is consequent to the first point, this book pre-empts a shift from mere ascription of urban vulnerabilities and challenges to an aggressive scrutiny of how those realities enable specific kinds of urban life to flourish. The anticipated theoretical heading theorized the postcolonial African city as "a huge intersection of bodies in need, and with desires ... [which] sustain themselves by imposing themselves in critical junctures, whether these junctures are discrete spaces, life events, or sites of consumption or production" (Simone 2004b, 3). In other words, the city's deepest truth is most evident in the human capacities to inhabit it, the manoeuvres that sustain such capacities, and the consequences of deploying such capacities. Edgar Pieterse suggests studies that can

> Potentially yield the microscopic details of everyday practices as imagined and experienced by the contemporary protagonists of the city who, through their abandonment by the nationalist development project, have been forced to carve out a distinctive, even if often monstrous, 'morality' of risk, chance, narcissistic pleasure and, also, tenderness and intimacy.
>
> Edgar Pieterse (2011, 12)

Parnell and Robinson (2012, 602) agree about the need for a "decentring [of] the dominant critique of urban neoliberalism" as a prerequisite to "activate the potential to creatively engage with urban development practice." Further, they note such a practice as a "generative site for urban theorization in its own right as well as a resource for wider scholarly reflection on the dynamics and experiences of cities across the global south." Jonathan Rigg (2007, 8) reiterates the "importance of the everyday and of grounded, micro-level perspectives" as the sole apertures that "can shed light on many of the critically important 'why' questions" about global south cities. Through what has been designated as crisis urbanism, this book joins these scholars in charting new theoretical frames to understand the complexity of Africa's postcolonial urbanism through urban film narratives.

Africa's postmillennial urban cinema, crisis urbanism, and urban futures

Within the general line of thought that postmillennial urban African cinema, as part of postcolonial intellectual and cultural traditions in Africa, is indeed a resource through which artists articulate emergent responses to urban crisis, this book has sought a localized engagement with alternative visions of Africa's urban futures. Eschewing the existing stereotypical metaphors of "third world cities" (Davis 2006), the book has explored these six African cities in the context of the third world as an interstice – a "third way or path between the capitalist and communist protagonists of the cold war" (Drakakis-Smith 2000, 1) – linked not only by the shared heritage of colonialism or legacy of poverty (Ibid., 2) but also through the sense a of a common futurity of crisis. The book has framed crisis urbanism in terms of everyday urban practices as in situ responses to an ever-present condition of crisis in postcolonial urban Africa. This has been achieved by using film narratives and visual language as exposé of the conceptualized responses to urban struggles. Such cinematic representations of crisis in various cities, it is surmised, offer up a protégé of contemporary epistemologies and postcolonial consciousness of crisis urbanism in Africa.

Largely, this is the challenge undertaken in this book: to propose that manoeuvres of coping with persistent crisis are not secondary to Africa's postcolonial urbanism but rather have become a dominant form of urbanism. To this end, the book took the reader through the emergence of nonentity in Johannesburg's townships as a form of post-apartheid urban crisis, the duress of *laissez faire* economic practices in Kinshasa, the crisis of inhabiting Nairobi as a (non)commonly shared city, the emergence of the rarray lifestyle as a predominant form of postwar urbanism in Monrovia, the latencies of crisis citizenship in Cairo, and the postwar cantonment in outlier urban lifestyles in Luanda. Through a critique of cinema representations of these ways of responding to present urban crisis, the book orientates the reader to the subtextual elements of urban crises and their postmillennial imaginary in postmillennial urban films. The films' narratives summarized below describe a maze of related urban narratives that do not avail themselves of any systematic logic but convey different forms of crisis urbanism from the perspective of contemporary urban life. Starting from an interdisciplinary position advocated in Chapter 1 and working within a qualitative framework, the book has shown this methodology of urban research as useful and productive. By working with postmillennial films to theorize Africa's postcolonial crises, this book has substantiated such an approach as a viable academic enterprise.

In Chapter 2, the book discussed the normalization of nonentity as a dominant form of urbanism among Johannesburg's post-apartheid townships. The chapter critiques the township as a set of urban practices embodied by Johannesburg's poor residents. Nonentity here represents a post-apartheid urban phenomenon, whereby township residents are not explicitly segregated, but remain the urban unknowns in terms of non-recognition within official welfare mechanisms and their own incapacity to fully integrate with post-apartheid urban life. It is argued that the idea of non-recognized citizenship is the new framing of citizenship crises in Johannesburg's postmillennial cinema. The chapter has engaged with two questions: if the township's past typified apartheid ideals, how do we make sense of a present and future in which official apartheid has been replaced by the rhetoric of the Rainbow Nation? By illustrating the outdated symbolism of the township, the chapter has demonstrated a need to look for new framings of the township, such as that can be found in urban film narratives, which evoke the characterization of city spaces through embodiment. On *Tsotsi's* representation of South African townships post-1994, the key argument is that through the use of noir cinematography and the film's mise-en-scène choices, the township appears as a simulacrum of nonentity. If we probe the future of the township, what metaphors of South Africa's post-apartheid urbanism does this analysis allow? The response conjectures the township as a set of urban practices embodied by Johannesburg's poor residents, and hence as a simulacrum of nonentity, that is, as an unintelligible and omitted place.

Related to the idea of nonentity is the emergence of *laissez faire* urbanism as a mainstay economic survival framework complete with attendant urban practices, as discussed in Chapter 3. Focusing on Kinshasa, this chapter critiques *laissez faire* urbanism as a form of postcolonial crisis urbanism. The key argument is that Kinshasa's dystopic conditions are embodied by popular everyday identity practices – the *romains*, *bana kwatas*, *chayeurs*, *gaddafis*, *chargeurs*, *cambistes*, and *mamas manoeuvre* – that exemplify the pervasiveness of bleak economic prospects as the mainstream form of urban economic practice. These practices are also emblematic of the realism of economic crisis within the city. The chapter has tended to the question: how do we interpret a city where livelihoods are not guaranteed, but rather achieved through speculation? The response, it is argued, demonstrates the paradox that is the reality between the imagined success of the *laissez faire* economic model and its catastrophic consequences. This is especially seen in la sapeur, a character that fakes urban opulence, thus personifying the unconventional responses to urban economic distress that have become a persistent reaction to crisis urbanism in Kinshasa.

Following ideas of *laissez faire* urbanism, Chapter 4 discusses the crisis of the urban commons in postcolonial Nairobi. That is, the struggle to survive through self-regulated economic manoeuvres can be compared to the unconventional spatial manoeuvres involved in sharing postcolonial space in Nairobi. Citing recent displacements in parts of the city, and juxtaposing these with the ongoing tussle around the urban real estate portfolio, this chapter hinges on Simone's (2007, 81) theory that "spaces full of garbage tend to be popularly perceived as outside municipal control, not subject to the authority or management of urban government." Seeing the provisional life typified by such trashed spaces as emblematic of the crisis of commoning in postcolonial Nairobi, the chapter argues that most Nairobians inhabit sites considered unfit for human habitation as the most viable approach to gaining inclusion to the city. Such occupation of "garbage sites," it is argued, designates a form of crisis urbanism that experiments with various tactics to share the city. The chapter makes two key arguments. First, that the aesthetic of garbage as signifying the difficulties of sharing the city suggest that Nairobi's downtown is integral to but fundamentally different from the mainstream city. Second, that from its signification of spatial marginality, we can see downtown as an inventory of the tensions of urban coexistence and the manoeuvres that become necessary to inhabit such a historically polarized colonial city as Nairobi. Both of these aspects are emblematic of the crisis of practicing urban commoning in postcolonial Nairobi.

Chapter 5 feeds into the previous conversation of problematic commoning in urban spaces. Focusing on urban residents' inability to achieve commoning in postwar Monrovia, it offers a unique perspective on how some have adopted an alienated urban, personified by the rarray character, life at once aligned with the war past as well as a new kind of postwar urban life. This chapter discusses the representations of the superficies of Monrovia's rarray characters as constituting a peculiar postwar urbanism. The chapter asserts that cinema imaginaries of a wartime Monrovia characterized by violent seizures of city spaces and the everyday life of Monrovia's residents in the aftermath of the war usefully emblematize the discursive role of the inertia of violence in producing crisis urbanism. The urban life of the rarray character – both a direct consequence of war and a lingering abstraction of its aftermath – thus demonstrates the emergent forms of urban life that have come to characterize Monrovia's crisis urbanism in the city's "postcolonial" times.

Closely related to alienated urban life is Cairo's crisis of urban citizenship, discussed in Chapter 6. This chapter builds a connection between the overt street protests that erupted in the city in 2011 and the representation of covert disenfranchisement as an urban condition that existed long before then. Through a survey of post-2011 urban film narratives and media reports, the chapter's main argument is that these representations convey an ongoing system of oppression and latent resistance. This covert revolution, more than its climax labelled the Egyptian revolution, conveys the crisis of inhabiting Cairo, and thus provides a useful meaning of urban citizenship in postcolonial Cairo. The chapter works with ideas such as immigration, passports, transport, freedom, and state services as frontiers.

Finally, the book discusses representations of Luanda's outlier urbanism, as based on postwar practices of the cantonment that typifies a form of crisis urbanism. Looking closely at the film *O Herói*, this chapter surfaces the film's underlying motif of crisis urbanism, which it indexes in diverse forms and contexts that demonstrate the reality of characters trapped within the hardships of peripheral urban experience. The chapter argues that confinement in wretched urban life is emblematic of the project of exclusive postcolonial urbanism in Luanda. Using cantonment and outlier urbanism as a conceptual lens, the chapter critiques the representation of Luanda's urban form and the protagonist's experiences in various urban sites as constituting a useful perspective on Luanda's postwar urbanism. That film characters' oscillation towards a better life is juxtaposed with a sense of despair, it is argued, portrays an elaborate scheme of alienation within the city. The theorization that follows is twofold. First, that urban characters coping with inescapable incapacitation in their pursuit of postcolonial inclusion symbolize a verisimilitude of postwar cantonment and second, that such urban restriction produces peculiar urban practices, theorized here as outlier urbanism.

Through these analytical prisms of crisis urbanism, this book situates postmillennial urban film narratives as the "critical epicentre of a politics of representation within the terrain of the global postcolonial world" (Bakari 2000, 11). These films, the book has argued, map out the forms and consequences of urban exclusion beyond the stereotypical theoretical paradigms of identity, economy, power, abandonment, governance, and fortification. Instead, the book understands these consequences as constituting emergent urban practices in postcolonial urban Africa, labelled as crisis urbanism, or the new common sense (Mbembe 2001) of emerging urbanism. Moreover, the fixation of such film narratives with verisimilitude of urban life draws on the postcolonial urge for self-narration and a desire to confront postcolonial urban peculiarities beyond the millennium. With the fixed forms of postcolonial citizenship experiences seen in these narratives anchored not so differently from those of the colonial past, the idea of the "post" (colonial, millennial) thus acquires more of a sense of stasis than transition. "Post" here thus designates the lack of transition, the wrong transition, or transition backward. It must be understood in the historical sense as a concept that does not undermine other related suffixes of colonial or millennial. It rather opens possibilities to talk of fixity, exacerbation, clones, or other not so "post-something" variations of colonial urban conditions that have been overlooked with the rise of global south theories and to posit these as the norm for Africa. Consequently, then, the book approaches crisis urbanism not just as a present condition of urban life in urban Africa today but as a condition that will continue to characterize many African cities in the future. Notwithstanding the possibility that these crises could change or be replaced by other forms of crises, crisis urbanism may register a prolonged presence. In sum, using crisis urbanism as a paradigm to support further theoretical work on postcolonial urban Africa, this book is a starting point intended to provoke further debates rather than offer fixed answers.

Bibliography

Appadurai, Arjun. 1996. *Modernity at Large: Cultural Dimensions of Globalization.* Minneapolis: University of Minnesota Press.

Bakari, Imruh. 2000. "Introduction: African Cinema and the Emergent Africa." In *Symbolic Narratives/African Cinema: Audiences, Theory and the Moving Image,* edited by June Givanni, 3–24. London: British Film Institute.

Beall, Jo, Owen Crankshaw, and Susan Parnell. 2002. *Social Differentiation and Urban Governance in Greater Soweto: A Case Study of Post-Apartheid Reconstruction*. Working Paper no. 11, London: Crisis States Programme, Development Research Centre.

Bradbury, Jill. 2012. "Narrative Possibilities of the Past for the Future: Nostalgia and Hope." *Peace and Conflict: Journal of Peace Psychology* 18 (3): 341–350.

Charlesworth, Esther, ed. 2005. *City Edge: Case Studies in Contemporary Urbanism*. Amsterdam: Elsevier.

Comaroff, Jean, and John L. Comaroff. 2014. *Theory from the South: How Euro-America Is Evolving Toward Africa*. Stellenbosch: SUN MeDIA.

Davis, Mike. 2006. *Planet of Slums*. London and New York: Verso.

De Boeck, Filip. 2011. "Spectral Kinshasa: Building the City Through an Architecture of Words." In *Urban Theory Beyond the West: A World of Cities*, edited by Edensor Tim and Mark Jayne, 309–326. London and New York: Routledge.

Drakakis-Smith, David. 2000. *Third World Cities Second Edition*. London and New York: Routledge.

Edjabe, Ntone, and Edgar Pieterse. 2010. *African Cities Reader*. Vlaeberg: Chimurenga and the African Centre for Cities.

Fredericks, Rosalind. 2018. *Garbage Citizenship: Vital Infrastructures of Labor in Dakar, Senegal*. Durham and London: Duke University Press.

Fujita, Kuniko, ed. 2013. *Cities and Crisis: New Critical Urban Theory*. London: SAGE Publications.

Gastrow, Vanya. 2018. *Problematizing the Foreign Shop: Justifications for Restricting the Migrant Spaza Sector in South Africa*. Samp Migration Policy Series No. 80, Cape Town: Southern African Migration Programme.

Gilroy, Paul. 1993. *The Black Atlantic Modernity and Double Consciousness*. London and New York: Verso.

Harvey, David. 2003. "The City as a Body Politic." In *Wounded Cities Destruction and Reconstruction in a Globalized World*, edited by Jane Schneider and Ida Susser, 25–46. Oxford: Berg.

Koonings, Kees, and Dirk Kruijt. 2009. *Megacities: The Politics of Urban Exclusion and Violence in the Global South*. London and New York: Zed Books.

Kruger, Loren. 2013. *Imagining the Edgy City: Writing, Performing, and Building Johannesburg*. New York: Oxford University Press.

Lyotard, Jean-François. 1984. *The Postmodern Condition: A Report on Knowledge*. Translated by Geoff Bennington and Brian Massumi. Manchester: Manchester University Press.

Madani-Pour, Ali. 1995. "Reading the City." In *Managing Cities: The New Urban Context*, edited by Patsy Healey, Stuart Cameron, Simini Davoudi, Stephen Graham and Ali Madani-Pour, 21–26. Chichester, West Sussex: John Wiley & Sons.

Mbembe, Achille. 2001. *On the Postcolony*. Berkeley/Los Angeles/London: University of California Press.

Mbembe, Achille, and Sarah Nuttall. 2004. "Writing the World from an African Metropolis." *Public Culture* 16 (3): 347–372.

Mbembe, Achille, and Sarah Nuttall. 2007. "Afropolis: From Johannesburg." *PMLA* 122 (1) *Special Topic: Cities*: 281–288.

Murray, Martin J. 2008. *Taming the Disorderly City: The Spatial Landscape of Johannesburg After Apartheid*. Ithaca, NY: Cornell University Press.

Mututa, Addamms Songe. 2019. "Johannesburg in Transition: Representing Street Encounters as Racial Registers in Clint Eastwood's Invictus." *Critical Arts South-North Cultural and Media Studies* 33 (2): 1–13.

Myers, Garth Andrew. 2005. *Disposable Cities: Garbage, Governance and Sustainable Development in Urban Africa*. Hampshire and Burlington: Ashgate Publishing Limited.

Myers, Garth. 2011. *African Cities: Alternative Visions of Urban Theory and Practice*. London and New York: Zed Books.

Nas, Peter J.M., Marlies de Groot, and Michelle Schut. 2011. "Introduction: Variety of Symbols." In *Cities Full of Symbols: A Theory of Urban Space and Culture*, edited by Peter J.M. Nas, 7–26. Amsterdam: P.J.M. Nas/Leiden University Press.

Neuwirth, Robert. 2005. *Shadow Cities: A Billion Squatters, a New Urban World*. London and New York: Routledge.

Oldfield, Sophie. 2014. "Critical Urbanism." In *The Routledge Handbook on Cities of the Global South*, edited by Susan Parnell and Sophie Oldfield, 7–8. London and New York: Routledge.

Parnell, Susan. 2014. "Conceptualizing the Built Environment: Accounting for Southern Urban Complexities." In *The Routledge Handbook on Cities of the Global South*, edited by Susan Parnell and Sophie Oldfield, 431–433. London and New York: Routledge.

Parnell, Susan, and Jennifer Robinson. 2012. "(Re)theorizing Cities from the Global South: Looking Beyond Neoliberalism." *Urban Geography* 33 (4): 593–617.

Pieterse, Edgar. 2008. *City Futures: Confronting the Crisis of Urban Development*. Lansdowne: UCT Press.

Pieterse, Edgar. 2011. "Grasping the Unknowable: Coming to Grips with African Urbanisms." *Social Dynamics: A Journal of African Studies* 37 (1): 5–23.

Pieterse, Edgar. 2014. "*Epistemological Practices of Southern Urbanism*." Draft Paper to be presented at *the ACC Academic Seminar*. University of Cape Town.

Pieterse, Edgar, and AbdouMaliq Simone. 2013. *Rogue Urbanism: Emergent African Cities*. Johannesburg: Jacana Publishers in association with African Centre for Cities.

Rigg, Jonathan. 2007. *An Everyday Geography of the Global South*. London and New York: Routledge.

Roy, Ananya. 2009. "The 21st-Century Metropolis: New Geographies of Theory." *Regional Studies* 43 (6): 819–830.

Simone, AbdouMaliq. 2004a. *For the City Yet to Come: Changing African Life in Four Cities*. Durham and London: Duke University Press.

Simone, AbdouMaliq. 2004b. "People as Infrastructure: Intersecting Fragments in Johannesburg." *Public Culture* 16 (3): 407–429.

Simone, AbdouMaliq. 2007. "Assembling Douala: Imagining Forms of Urban Sociality." In *Urban Imaginaries: Locating the Modern City*, edited by Alev Çınar and Thomas Bender, 79–99. Minneapolis: University of Minnesota Press.

Smith, Neil. 1996. *The New Urban Frontier: Gentrification and the Revanchist City*. London and New York: Routledge.

Taylor, Peter J. 2004. *World City Network: A Global Urban Analysis*. London: Routledge.

United Nations Population Fund. 2007. *State of the World Population 2007*. New York: UNFPA.

Index

Aasef ala el-iz'ag (2008) 98
Abaza, Mona 101
The Advocates for Human Rights
(2009) 86
African Centre for Cities (ACC) 2
Akkerman, Abraham 70
Ali, Mohammed 52
Al Jazeera 103
Ambe, Hilarious 77
amenities 61
Americo-Liberian invasion 91
Anon (2018) 84
anthropological/sociological data-based
studies 130
Arab cinema 99
Arab News (2017) 102
Arab Spring 15, 96, 97, 101
Asal Aswad (Molasses) (2010) 15, 96;
airport immigration desk 104–105;
Egyptian and American citizenship
98–99, 103–104; localizing inequalities
104; public housing 106; social,
political and cultural contexts 99;
undercover resistance 106–109;
urban mobility 106; urban pressures 100; vacant nationalism
99–100; visitation rights and oppression 105
"asymmetrical relations of power," 117

Bakupa-Kanyinda, Balufu 41
Bamako (2006) 9
Batepá (2010) 114
Battle for Johannesburg (2010) 22
Beasts of No Nation (2015) 77, 79–80
Bender, Thomas 4
Best Foreign Film Academy Award,
2006 25
Blood Diamond (2006) 82
Body Team 12 (2015) 83
Bolbol Hayran (2010) 98

Bradshaw, Peter 84
Brown, Julian 24

Cairo's crisis of urban citizenship 15,
134; *Asal Aswad* 98–100; explicit
performances of revolt 97; indignity
105–109; latent repression 102–105;
mechanisms of control 97; occupation
97; revolt under construction
100–102; social dualism 97; street
protest 96–97; tactical manoeuvres 97;
undercover resistance 105–109;
violence use 97–98
Camp 72 (2015) 83
Cape of Good Hope (2004) 22
Cape Town World Cinema Festival 114
Castells, Manuel 4, 104
censorship 102, 105
Chahine, Youssef 102
Chang, Chris 41
Chaos (2007) 102
Charter of National Action 105
Chikin Biznis–The Whole Story
(1999) 22
Christensen, Maya M. 80, 89
"cinematic trope of blackness," 26
City of Joy (1992) 41
Çınar, Alev 4
Comaroff, Jean 9
Comaroff, John L. 9
*Comboio da Canhoca (Train of
Canhoca)* (1989) 114
conflict in cities 8
Congo in Four Acts (2010) 41
Conversations on a Sunday Afternoon
(2005) 22
Cook, Pam 124
Cornfeld, Ariela 70
Covid-19 lockdown 61
Cowherd, Robert 100
crime culture 23, 31

crisis urbanism 5–8
cyclic violence 85

Da Silva, Maria 24
De Boeck, Filip 4, 39, 43, 49
de Certeau, Michel 100–101
de Groot, Marlies 3
De Satge, Richard 2
"dialectics of contemporary world
   history," 5
digital video technologies 10–11
Diouf, Mamadou 4
District 9 (2009) 22
Doe, Samuel Kanyon 82, 87, 91, 92
dos Santos, Jose Eduardo 119, 126
Dovey, Lindiwe 22
Drakakis-Smith, David 131
dual citizenship 104
Duck, Leigh Anne 2

Ebola pandemic 83
Eckardt, Frank 7
Egyptian uprising 2011 98
18 Days (2011) 102–103
El-Hawary, Nouran Al-Anwar 99
Ellapen, Jordache Abner 25
el Magd, Nadia 98
El Said, Tamer 102
The Emigrant (1994) 102
Epistemological Practices of Southern
   Urbanism 2
European Programme for Sustainable
   Urban Development's 2010 report 7, 8

film narratives 10–12, 15, 82, 99,
   108–109, 130, 133–135
financial crisis 7–8, 47
Foreman, George 52
formal capital system 61
Foster, Dulce 87
Frassinelli, Pier Paolo 41
fraudulent economic networks 61
Fredericks, Rosalind 60
Freetown (2015) 77, 82
Fujita, Kuniko 7

Ganga, Maria João 114
Gattaca (1997) 84
Gazlam 22
Geenen, Kristien 43
Ghannam, Farha 106
globalization 24
global north-global south urban theory
   131–132; cross-disciplinary engagement 5–6; infrastructure and development 3; interpretation of cities 2–3;

people as infrastructure 4; research
   practice 2; settlements, economies, and
   culture 3–4; sociological and anthro-
   pological inquiry 4; urban space 5
God Is African (2001) 22
Goldman, Michael 97
Good Kill (2014) 84
Graceland (2004) 25
Graham, Meghan 82
Gulema, Shimelis Bonsa 105

Hagen, Erica 71
Hamadi, Dieudo 41
Hamid, Rahul 85
Harrow, Kenneth W. 118, 125
Hatred (2012) 83
*Heen Maysara (Waiting for Better
   Times)* (2007) 99
Hijack Stories (2000) 22
Hirson, Baruch 24
Hondros (2017) 82
The Host (2013) 84
housing projects 119, 122, 125
Human Rights Watch (2003) 117

Independent National Patriotic Front of
   Liberia (INPFL) rebels 91, 92
informal economy 12, 42–43, 47
informal habitation 14
informal settlements 14, 58, 60, 118
In the Last Days of the City (2016) 102
In the Shadow of Ebola (2014) 83
In Time (2011) 84
Invictus (2009) 23
I Want to be a Pilot (2006) 68

Jaguar, Charles 61
Jerusalema: Gangsters' Paradise (2008) 23
Johannesburg's post-apartheid townships 132, 133; modular narrative
   space 31–35; new urban imaginaries
   in cinema 22–23; unintelligible
   township 29–31; urban South Africa's
   future theory 23–26
Johnny Chien Méchant 2002 83–84
Johnny Mad Dog (2008) 15, 76, 83 85,
   89–92
Jump the Gun (1997) 22
"justifications and rationales," 80

Kabila, Laurent 44
Katembo, Kiripi 41
Kefaya 99
Kibera Kid (2006) 68–69
Kichwateli (2012) 65–68

Kinshasa Makambo (2018) 41
Koonings, Kees 2, 101, 130
Kruger, Loren 24
Kruijt, Dirk 2, 101, 130

Laaf Wa Dawaraan (2016) 98
laissez faire urbanism 13–14, 132, 133;
   acute fuel shortages 45; economic
   difficulties 42; exploitative relations
   45; individual-based survival 43–45;
   informal street economies 41–42;
   public mobility infrastructure 43–44;
   self-regulated economic practices
   42–43; survival, prolonged poverty,
   and illiteracy 44
La Nouvelle Liberte 59
la sapeur 50–53
The Last Face (2016) 83
Léfebvre, Henri 4, 23, 61, 100
Liberia: An Uncivil War (2004) 83
Liberia cinema 83
Liberian Girl (2011) 83
Liberians United for Reconciliation and
   Democracy (LURD) 83–84, 89, 90
Liberia's Civil War of 1989–2003, 83
Löfgren, Orvar 73
Lord of War (2005) 15, 76, 83–89
Luanda's postwar urbanism 15–16,
   134–135; *Comboio da Canhoca (Train
   of Canhoca)* (1989) 114; film produc-
   tion 114; *Na Cidade Vazia (The
   Hollow City)* (2004) 114–115; *O Herói*
   115–126
Lusala, Divita Wa 41

Madani-Pour, Ali 5
Makhubu, Mbuyisa 22
Mansour, Dina 99
Manuel Perlo Cohen's 2011 study 7–8
Marie, Khaled 98
markets 61, 83, 123
Maron, Mikel 71
Max and Mona (2004) 22
Mbembe, Achille 4, 50, 52, 86
McNamara, Joshua 66
Memória de um Dia (Memory of a Day)
   (1982) 114
Miners Shot Down (2014) 22
Mobutu Sese Seko 44
Movement for Democracy in Liberia's
   (MODEL) 89
MTV Movie Awards 41
Mubarak, Hosni 99
musseques 118, 119, 125
Myers, Garth Andrew 130

Na Cidade Vazia (The Hollow City)
   (2004) 114, 117
Nahr, Dominic 103
Nairobi Half Life (2012) 14, 58; access
   to city 64; communality 73; criminal
   gangs 65, 70, 71; economic opportunities 64; images of garbage 69–71;
   narrative valence 65–66; residents
   characterization 63, 72–73; spatial
   marginality 71–72; visual aesthetics 72
Nairobi's urban commons 133–134;
   Covid-19 lockdown 61; demolitions
   of city spaces 61–62; *Nairobi Half
   Life* 63–65, 69–73; prevalence 62; real
   estate portfolio acquisition 62;
   representations of garbage in cinema
   65–69; social community 62–63;
   spatial interests 63; trash problem
   58–61; zero-sum crisis 63
Nas, Peter J.M. 3
National Patriotic Reconstruction
   Assembly Government (NPRAG) 92
natural disasters 8
necropolis 15, 81, 86–89
Neto, António Agostinho 126
Neuwirth, Robert 59
Newman, Kim 85
Ngangura, Mwezé 41, 51
nonentity 21
nonentity theory 24
Nuttall, Sarah 24

O Herói 16; cantonment 117–121;
   outlier urbanism 121–126; representa-
   tions 116–117; wartime images 116;
   Zézé Gamboa's interview 115–116
Oldfield, Sophie 2, 4
Oppong, Joseph R. 45
*O Sol Ainda Brilha (The Sun Still
   Shines)* (1995) 114

Page, Thomas 83
Paris Film Festival 114
Parnell, Susan 2, 4–6
Peck, Raoul 41
Peel, Lilly 122
People and Power 103
People's Redemption Council
   (PRC) 82
Pièces d'identités (1997) 51
Pieterse, Edgar 2, 4, 24, 25, 131
Pieterson, Antionette 22
Pieterson, Hector 22
Plissart, Marie-Francoise 39, 43
political conflicts 24, 90

political revolution 98
population displacement to cities 8
postcolonial Kinshasa: economies of
    dystopia 45–47; Khadaffi 47–50;
    laissez faire urbanism 41–45; la sapeur
    50–53; post-death procedures 39;
    survival/recognition 39–40; *Viva Riva!*
    39–41
postwar Monrovia 134; Africa's civil
    war cinema 76–77; fractures 89–93;
    melancholy of dismissal 78–79;
    necropolis 86–89; postwar hustle
    79–80; rarray urbanism 80–85
Pray the Devil Back to Hell (2008) 82–83
"primitive accumulation," 73
private equity 61
Proteus (2003) 22
Publishers Weekly (2005) 84

*Quem Faz Correr Quim (Who makes
    Quim Run?)* (1991) 114

racial segregation 24
Rape for Who I Am (2005) 22
rarray urbanism 14–15; expendability
    and marginality 82; Monrovia's
    history in cinema 82–85; political and
    economic benefits 81; unemployed
    youths 80–82
"re-baptism," 46
Rebelo, Rudi 115
Rigg, Jonathan 131
rights to the city 61
Robinson, Jennifer 5–6
Rodrigues, Cristina Udelsmann 122, 126
Rogerson, Christian M. 24
Roy, Ananya 129

Saied, Shaimaa 98
Sauvaire, Jean-Stephane 83
Schut, Michelle 3
shabāb al-thawra 98, 100
Simone (2002) 84
Simone, AbdouMaliq 4, 24
Sirleaf, Ellen Johnson 82
Small Small Thing (2013) 83
Smith, Adam 43
social dualism 97, 100
social inclusion/exclusion 120
social justice 98
social media journalism 102
socio-economic mobility 120–121, 125
The Sparrow (1973) 102
spatial memory 32
Stockton, R.F. 90–91
Sumégné, Joseph-Francis 59

Tabishat, Mohammed 102
Tamantashar Yom (2011) 98
Taylor, Charles 84
Taymour and Shafika (2007) 98
Tengers (2007) 22
This is Congo (2017) 41
Tolbert, William 82
Trapido, Joe 42
Trefon, Theodore 46
True Whig Party (TWP) in 1980 82
Tsha Tsha 22
Tsotsi (2005) 13, 25–26; embodiment of
    nonentity 27–29; township as a
    modular space 31–35; visualizing the
    unintelligible township 29–31
2008 global economic crisis 7–8

UNITA soldiers 117
*Un lugar limpio y bien iluminado (A
    Clean and Well-Lit Place)* (1991) 114
urban commons 62–63
urban crisis contexts: degenerative
    urbanism 1; economic and financial
    crisis 7–8; global north-global south
    urban theory 2–6; humanitarian crises
    8; imaginary and representation 9–10;
    in situ practices 8–9; perspectives 1–2;
    postmillennial films 10–12; Rome's
    judicial institutions, politics, and
    political economy 7; unconventional/
    illegal strategies 6–7
urban identity 13, 21, 31, 77, 78, 82
urban inequality 59–60, 116
urban knowledge 2
"Urban Nostalgia: Colonial traces in
    the postcolonial city of Luanda," 124
urban theorization 4, 6, 129, 131
Utas, Mats 80, 89

Viva Riva! (2010) 14; *see also* postcolo-
    nial Kinshasa

Watson, Vanessa 2
Welcome to Our Hillbrow (2001) 25
West-Pavlov, Russell 25
White, Bob W. 42
Whiteley, Gillian 59
Wickham, Chris 7
Wits Institute for Social and Economic
    Research (WISER) 124
Wooden Camera (2003) 22
Woodruff, Tanya 45

Yizo Yizo 22

Zulu Love Letter (2004) 22

Printed in the United States
by Baker & Taylor Publisher Services